THE THEORY OF EVERYTHING ALL THE WAY UP

Enhanced Print

Austin P. Torney

Copyright

To the Depths

Nought   Naba

Zero   Nothing

Of the Deep

To the Depths of the Deep
(After Shelley)

Here I stand, holding fast.
Onto my other half.

The zephyr faints, dying in the half-light,
Its caress suspended, as day kisses night,
When, for some instants, stretching into moments,
We are neither here nor there, but in twilight.

We live at this boundary of day and night,
Our selves merging in the blend of twilight:
You and me, me and you; yours, mine, and ours—
The day-gold melts into the jeweled night.

Above us, fires burn the stars away;
Below us, the Earth turns under our feet;
Within us, unworded dreams haunt the soul;
Around us, night pours blackness on the ground.

Soft and warm, the evening caresses us,
In gentle darkness, and quiet stillness.

Here we sense the sweep across the heartstrings,
For we're undistracted by the day's bright noise.

I beg of the night to yield its dearest puzzle,
To reveal the full truth of what it is.

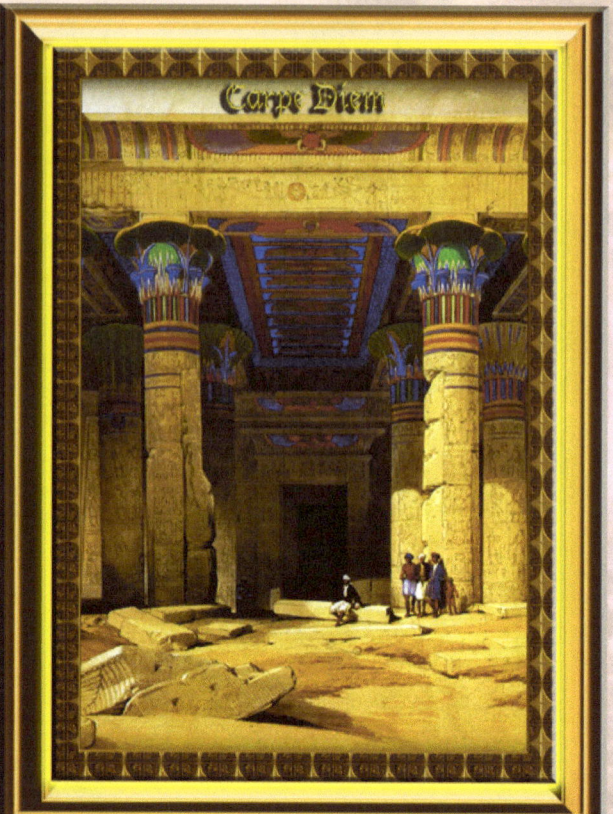

Much I already know from twilight dreams,
And from poems unveiling truth and beauty,
But, I ask, with my most inquiring looks
To know the deepest secrets of the night.

So, I must ask from the powers of the night,
Not immortality, nor youth, nor birth,
But only that I glimpse the enigmatic:
That riddle solved of the conundrum.

The door resisted at first,
Then creaked into the crypt,
Powdered rust streaming from the hinges.

Here the answer to All was kept,
But not all was pleasant—it spoke of death,
Of life's end, separate by just a breath...

I saw tombstones overgrown, underswept,
Names unknown—and to all the message saith:

— "Read Me" —

It said, in words engraved beyond the brink—
"You, who live, up above: of life go drink;
And you, underneath, now lying so dead:
Rest in peace, RELAX—it's later than you think!"

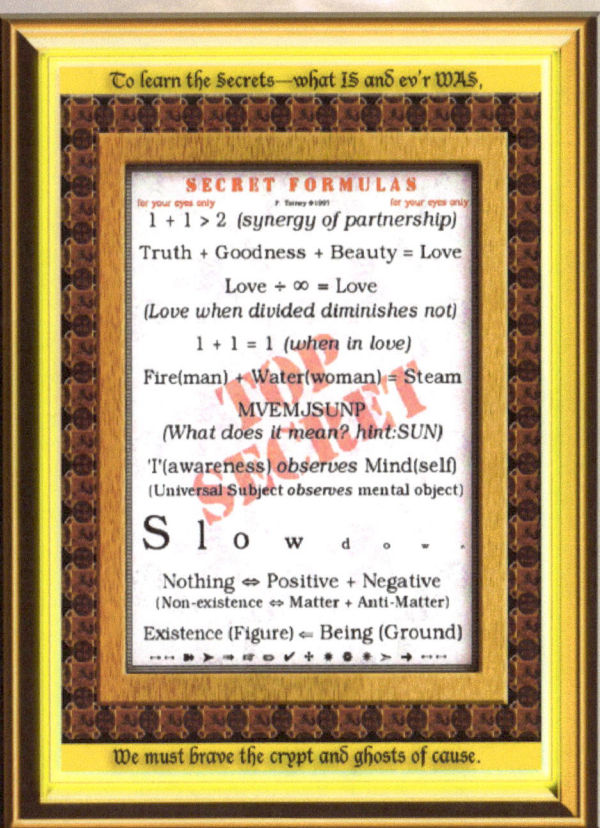

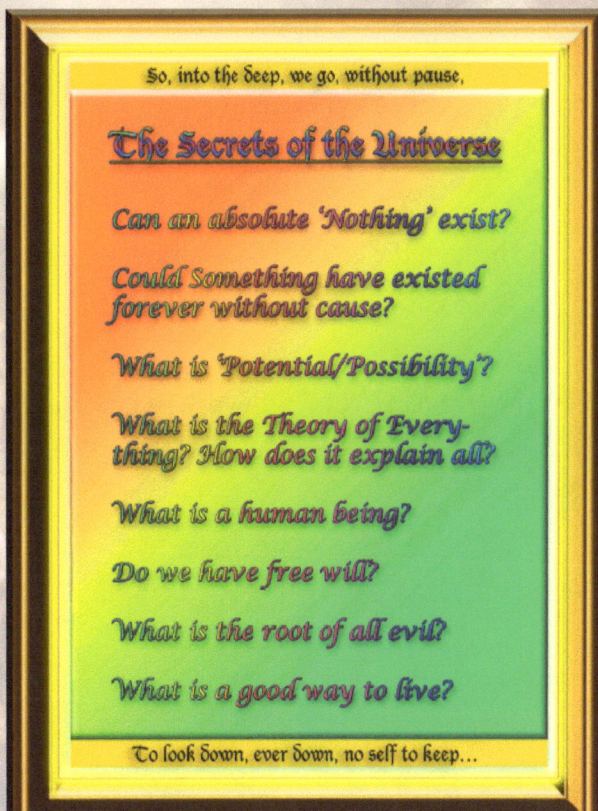

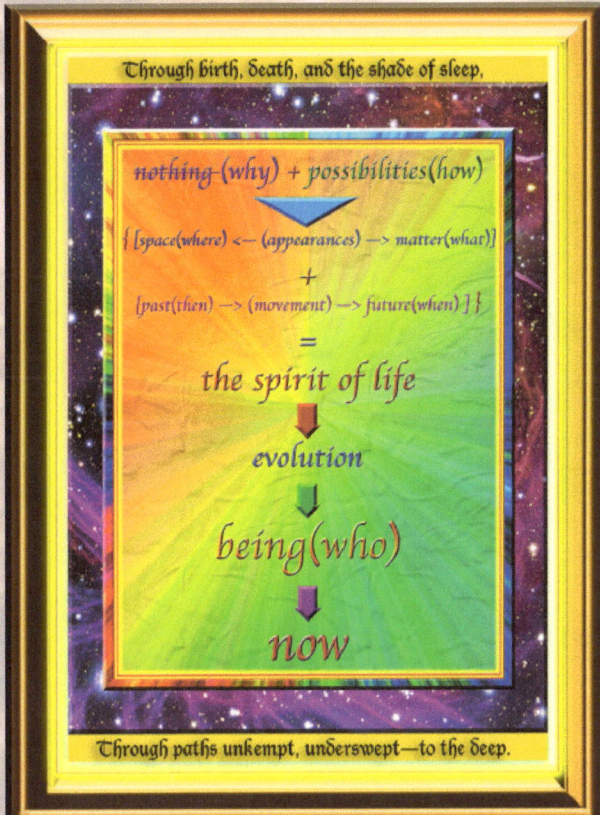

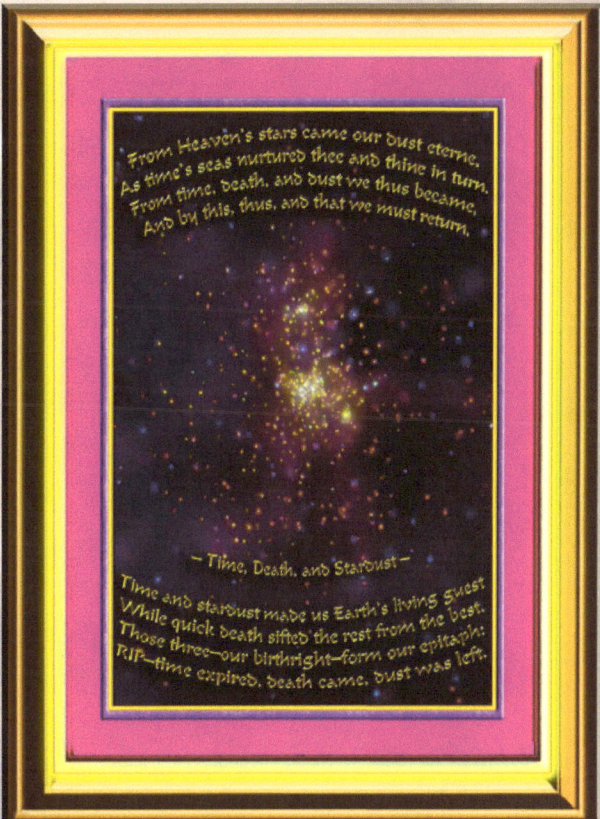

Panel 1 (top left):

To learn the Secrets—what IS and ev'r WAS,

SECRET FORMULAS

for your eyes only ᵀ Tierney ®1997 for your eyes only

1 + 1 > 2 (synergy of partnership)

Truth + Goodness + Beauty = Love

$$Love \div \infty = Love$$
(Love when divided diminishes not)

1 + 1 = 1 (when in love)

Fire(man) + Water(woman) = Steam

MVEMJSUNP
(What does it mean? hint: SUN)

'I'(awareness) observes Mind(self)
(Universal Subject observes mental object)

S l o w   d o w .

Nothing ⇔ Positive + Negative
(Non-existence ⇔ Matter + Anti-Matter)

Existence (Figure) ⇐ Being (Ground)

TOP SECRET

We must brave the crypt and ghosts of cause.

To learn the Secrets—what IS and ev'r WAS,
One must brave the crypt and ghost of cause...

Panel 2 (top right):

So, into the deep, we go, without pause,

The Secrets of the Universe

Can an absolute 'Nothing' exist?

Could Something have existed forever without cause?

What is 'Potential/Possibility'?

What is the Theory of Everything? How does it explain all?

What is a human being?

Do we have free will?

What is the root of all evil?

What is a good way to live?

To look down, ever down, no self to keep...

So, into the deep, we go, without pause,
To look down, ever down, no self to keep—

Panel 3 (bottom left):

Through birth, death, and the shade of sleep,

nothing(why) + possibilities(how)

{ [space(where) ⟵ (appearances) ⟶ matter(what)]

+

[past(then) ⟶ (movement) ⟶ future(when)] }

=

the spirit of life

evolution

being(who)

now

Through paths unkempt, underswept—to the deep.

Through birth, death, and the shade of sleep,
Through paths unkempt, underswept—

Panel 4 (bottom right):

From Heaven's stars came our dust eterne,
As time's seas nurtured thee and thine in turn,
From time, death, and dust we thus became,
And by this, thus, and that we must return.

— Time, Death, and Stardust —

Time and stardust made us Earth's living guest
While quick death sifted the rest from the best.
Those three—our birthright—form our epitaph:
RIP—time expired, death came, dust was left.

To the deep,
Through the cloudy strife
Of this hazy life,
Through the equations of eternity—
Their non-paternity nor maternity,

Past the realm of the things which seem or are,
Even o'er the steps of the remotest bar.

Down, down!

Where the mind whirls round and round,
As the ear draws forth the sound,
As the eye sees the light,
And of the dark the fright.

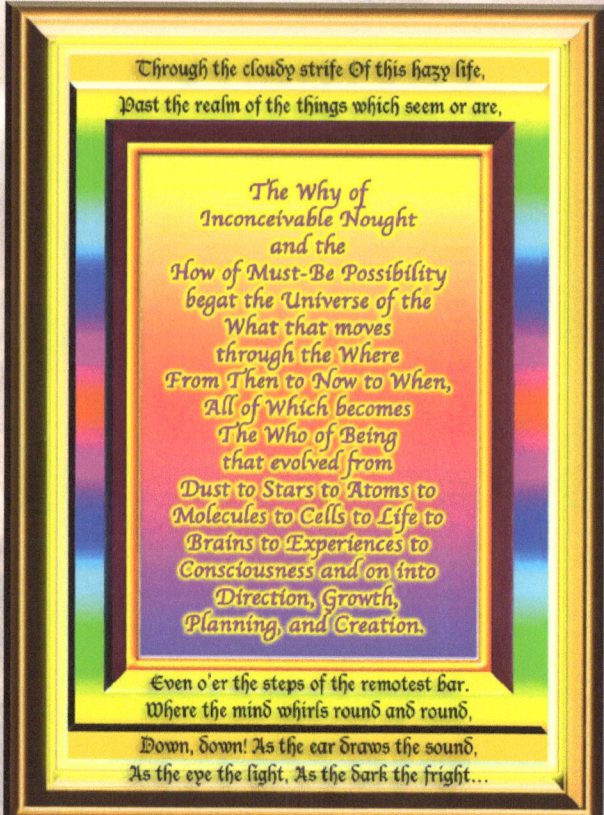

Through the cloudy strife Of this hazy life,
Past the realm of the things which seem or are,

The Why of
Inconceivable Nought
and the
How of Must-Be Possibility
begat the Universe of the
What that moves
through the Where
From Then to Now to When,
All of Which becomes
The Who of Being
that evolved from
Dust to Stars to Atoms to
Molecules to Cells to Life to
Brains to Experiences to
Consciousness and on into
Direction, Growth,
Planning, and Creation.

Even o'er the steps of the remotest bar.
Where the mind whirls round and round,
Down, down! As the ear draws the sound,
As the eye the light, As the dark the fright…

Beyond all death, despair, love, and sorrow,

Secrets of the Universe

Past yesterday, today, and tomorrow…

The body's guide is but the spirit of the mind.

Stars generate the lower elements;

Supernovae generate the higher ones…

Atoms form the molecules that lead to

Life's complexity—from simplicity.

Down, down! Through the fog, the not, and the void.

Down, down,
Beyond all death, despair, love, and sorrow,
Past yesterday, today, and tomorrow—

The body's guide is but the logic of the mind.
Down through the fog, the not, and the void,
Where "God" and everything fail; Oh, zoids!

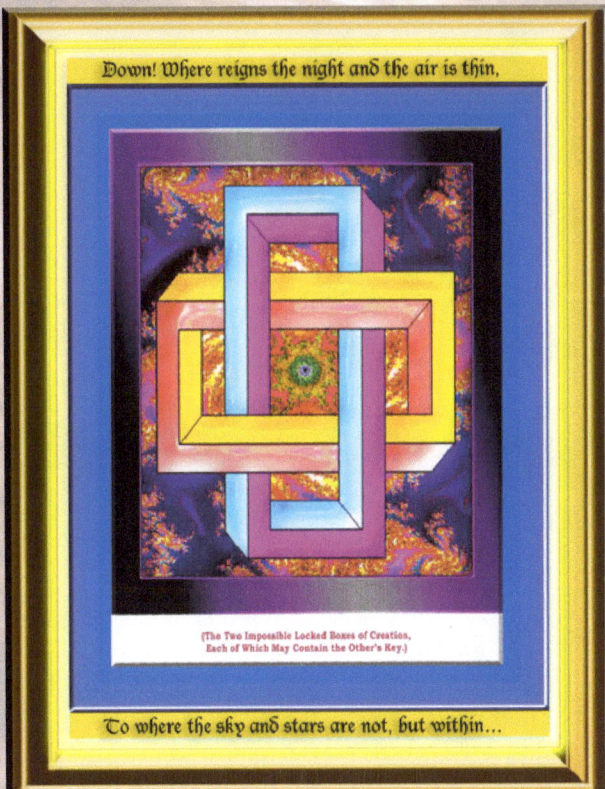

Down! Where reigns the night and the air is thin,

(The Two Impossible Locked Boxes of Creation, Each of Which May Contain the Other's Key.)

To where the sky and stars are not, but within…

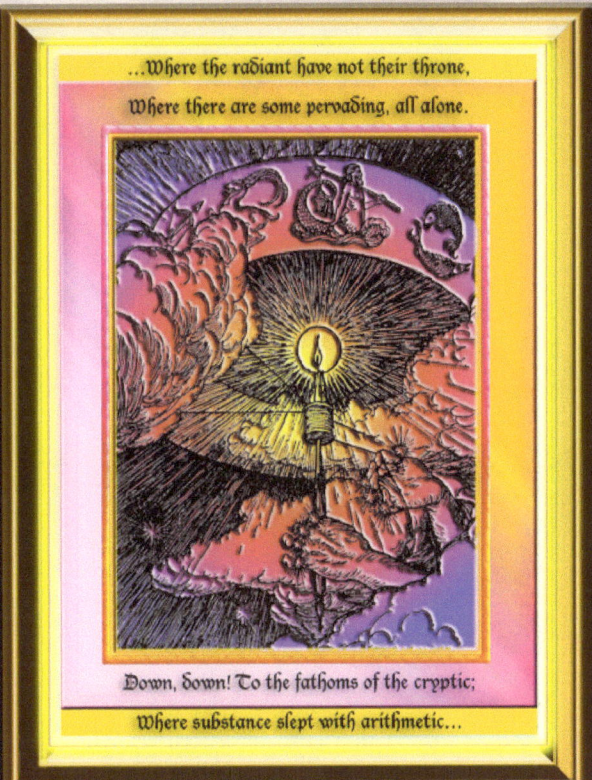

…Where the radiant have not their throne,
Where there are some pervading, all alone.

Down, down! To the fathoms of the cryptic;
Where substance slept with arithmetic…

Down!
Where reigns the night, and the air is thin,
To where the sky and stars are not, but within,

Where the radiant have not their throne,
Where there is one pervading, all alone.
Down, down!
To the fathoms of the cryptic;
Where substance slept with arithmetic,

Toward the spark yet nursed by embers,
To the first and last the universe remembers—
To seek the gem that shines—
The wealth of mines,
The jewels so treasured by thee and thine.

What truth accelerates life's momentous gem,
Letting the motto become "Carpe diem"?

Who seized the moment or lost its momentum,
Wearing not the time as its royal diadem?

The Secrets of the Universe

…Toward the spark yet nursed by embers,
To the first and last that the universe remembers,
To seek the gem that shines—the wealth of mines,
The jewels so treasured by thee and thine.

Down, down! We guide thee, we must carry thee;
Down, down! We're illumination beside thee…

Fear not the proof—it's the beauty of truth.

The World does not pass by—
We pass through it live;
Clear your being so the treasure may arrive;
This spirit sparkles of a different light—
The gemstones are of a different mine.

Down, down!
We guide thee, we must carry thee;
We're illumination beside thee...

Fear not the proof—
It's the beauty of the truth:

Above the ground, you were ever born again
When the roseate hearts were cleansed by dew;
And lucky were you if spring found you new,
As every blossom on the bush blew full.

When these wonders the new morning bestrew;
The beauty of truth was all that you "knew"...

Life's hardships there were softened by beauty,
All its weaknesses strengthened by the truth—
As, when roses blossomed, like realizations,
Beauty itself bloomed from the well of truth.

For now, rarely enough, existence is left aside,
And, yet, the essence ever has its other side—
Life, although anguishing, must be lived fully,
Since, if we're alive enough to feel its beauty
Then we're exposed to the opposite twin;
Yes, Beauty's other side is Melancholy.

Down, down,
The essence beckons us back home,
As like the contained-container is the poem.

— The Beauty of Truth —
Life's hardships can be softened by beauty,
Its weaknesses can be strengthened by truth.
When roses blossom, like realizations,
Beauty itself blooms from the well of truth.

When a deep truth is known so intensely
That all of its clothing falls away,
Then we have learned the beauty of truth, for
The reality of meaning is beauty.
Life, although anguishing, must be lived fully,
Since, if we're alive enough to feel its beauty,
Then we're exposed to the opposite twin—
Yes, Beauty's other side is Melancholy.

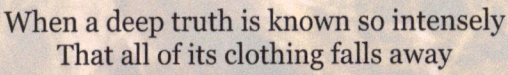

When a deep truth is known so intensely
That all of its clothing falls away

Then we have learned the beauty of truth,
For the reality of meaning is beauty.

When sadness brooded over the morrow,
I once visited the deep well of sorrow.
There enshrined, inseparate, Beauty said,
"'Twas from me that sadness you borrowed."

So, do we live the life of art,
Each playing our part?

Nay, that is not life, nor a part, bit,
For there's another dimension to it.
Art and poetry enrich human experience,
But they're no substitutes for the living of it.

Like Keat's figures on the urn, blest,
Should we live life any less?
NO!—because what is deathless is also lifeless!

Down, down!

Truth and beauty must be inseparable,
Although this is seemingly imponderable.

On that sphere above,
Soft breezes ever blew,
Caressing me and you,
As we kissed the roses new
And drank their dew.

Reason and passion then merged into one,
As truth and beauty made their rendezvous.

Down, down, ever down—

Through the antiquity, past all of the known—
Arriving at the lowest, remotest throne,
One of the highest perfection,
For it is of the opposite directions.

Opposite twins rule the causing call,
The positive and negatives being the All.

Here, the enigma of the ever immortal
Is undone and unloosed, through its portal—
The Theory of Everything mortal—
The Idea for which we've opened the door to.

Down, down,
To the end at last!

Here the lawless and the formless
Of the unordered, uncreated scene;
Here the causeless reigns supreme.

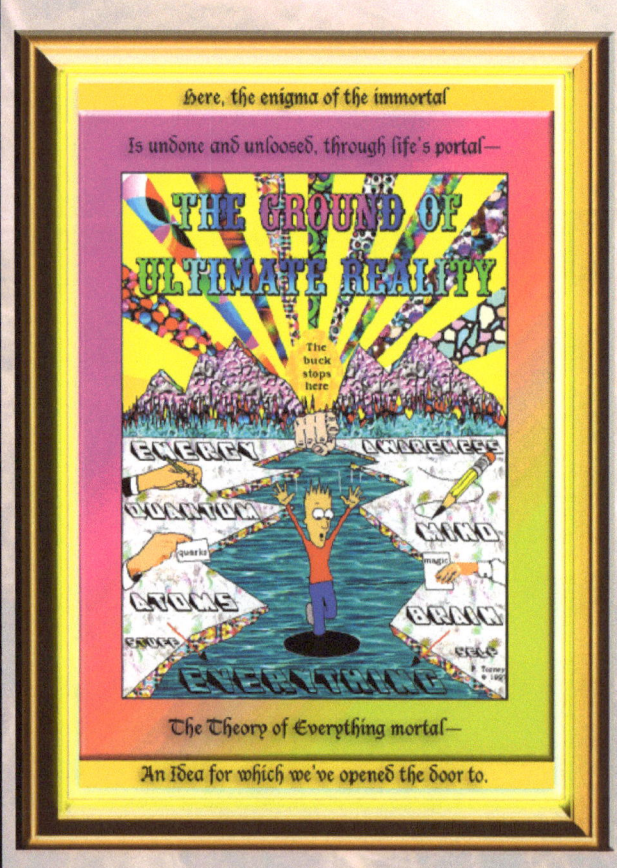

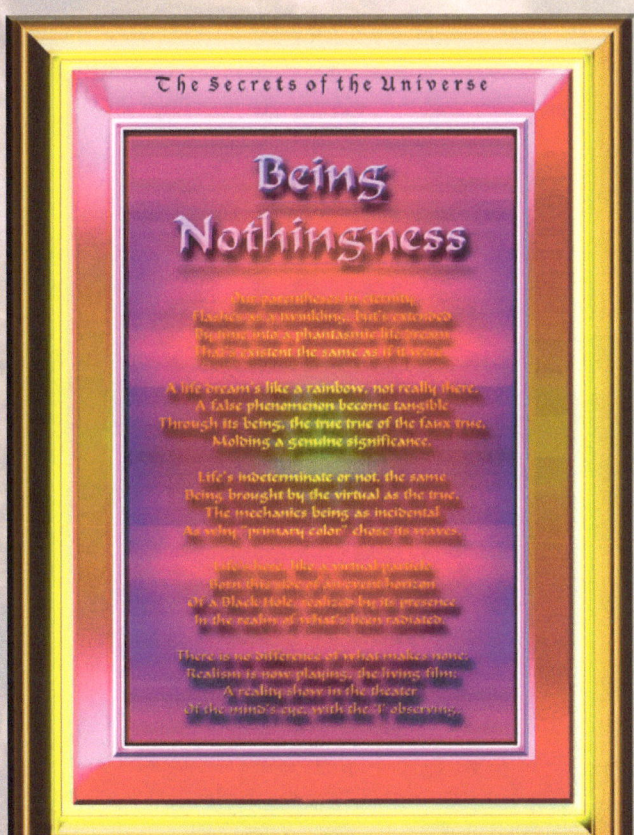

The Secrets of the Universe

Being Nothingness

Our parenthesis in eternity,
Flashes as unending, but extended
Incommensurate a phantasmic life means
That existent the same as if universal.

A life dream's like a rainbow, not really there,
A false phenomenon become tangible.
Through its being, the true true of the faux true,
Molding a genuine significance.

Life's indeterminate or not, the same
Being brought by the virtual as the true.
The mechanics being as incidental
As why "primary color" chose its waves.

Existence, like a virtual particle,
Born from the depths of uncorrected,
Of a Black Hole realized by its presence
In the reality of what's been radiated.

There is no difference of what makes none.
Realism is now playing, the living film;
A reality show in the theater
Of the mind's eye, with the 'I' observing.

Movie To the Depths.1 Part 2 of the Introductory Media

HD video part 1:

http://www.youtube.com/watch?v=VBt2_cUh-zQ

HD video part 2:

http://www.youtube.com/watch?v=HxpopjeLzlI

Everything

Such we are stirred, so touched by the starlight,

That it seems we'll ne'er be the same again.

Do we sense the euphony of the spheres?

Can we fathom the theory of everything?

Since we all become of this universe,

Should we not ask who we are, whence we come?

Insight clefts night's skirt with its radiance:

The Theory of Everything shines through!

The TOE

SUMMARY

1. There is nothing to make anything of, and this is for sure. No one can suggest any other source, for that would just be another thing, so, therefore, it is impossible for there to be any other source.

2. Contrast this sure thing to the apparently mistaken notion that a lack of anything couldn't do anything, that such a state of affairs would be a stable situation. It isn't, and the proof: there is something. In fact, we find nothing nowhere, with something now here.

3. There are no paradoxes, only misunderstandings.

4. The analysis is inviolate. There is literally nothing to make anything of.

5. The beginning, the middle, and the end are all of the same.

6. Eternity, Infinity, Everything, and Nothing are all of a package—the zero-sum balance of opposites.

The Theory of Everything (TOE) Summary

The TOE must not only encompass the unification of the forces, which is the GUT (the Grand Unification Theory), but it must also demonstrate why anything exists at all, and how it does so, and, furthermore, how that ties in to what we have become, up to now, such as how we operate, and so forth.

"Why does anything exist at all?" is much akin to the 'great' philosophical question of "Why is there something instead of nothing?" Both questions are stated backwards, as if 'something' had to be decided, versus there being Nothing, a total lack of anything. Some presume that something came from an absolute Nothing or that things are forever made of even smaller things. These are not the case, for causes cannot forever precede causes—an infinite regress, nor can a total Nothing ever produce anything, for 'it' has no existence, meaning that it is not even 'there', yet, Nothing still plays a role, as there is literally nothing to make anything of. To confirm, we look for a balance of opposites, summing to zero, and, indeed, we find it. It is as if Nothing is perfectly unstable, which is why it cannot be, and is not found anywhere at all.

So, 'something' is, because Nothing cannot be or stay?

Yes, the causeless 'something' must then be the normal and natural state of affairs, even if it's some things generated and is not always the same, exact something, for,

If there were a total lack of 'something' then this would still be the case, as there would continue to be no existence.

A chain of ever caused 'somethings' from lessor things would be an infinite regress.

Nothing, obviously, can't exist. See it? No, for it's not there.

So, what is this 'something'?

It is simple 'thing', for the complex ever always has simpler parts from which it is constituted, as we descend downward.

What is as simple as it gets?

Nothing, but that is not there.

Then what is the next simplest 'thing'?

Well, it must be a near 'nothing'.

Was 'it' designed or reasoned forth?

No, as it had to be causeless.

The buck has to stop somewhere.

So, we don't expect a blueprint of it, since it's an undefined chaos of disorder.That Is, of course, hard to swallow, as we always wish for definition; however, the unordered must be the answer, for there was nothing prior to it to order it or to reason/design it forth.

So, then, 'something' exists because there is no alternative?

Yes, and that is the 'Why' of existence, for it is the natural state, plus that, since it has no possible source, as in no cause for its effect, from any other thing, it must be of no-thing, Nothing, summing to it, overall.

We do note that things, and especially much simpler things readily combine, into more and higher composite complexities.

As usual, like all simpler things, the near 'nothing' is mostly unstable, too, for simple things ever go through phase changes, and many readily combine into the more complex. This would even ap-

ply to a total Nothing if it could even 'try' to be, for an instant, but it can't.

How do we know that the 'something' is a near 'nothing', or even zero overall?

The negative potential energy of gravity cancels out the positive kinetic energy of matter, leaving only the near 'nothing'.

Also, charge and angular momentum add to zero, or near zero, due to the quantum uncertainty/ fluctuations, and, more mainline, energy, too, is conserved.

3) Matter and anti-matter particles are always produced in pairs, with opposite polarity of charge, such as electrons and positrons.

4) A zero-sum equation could be at work, replacing the non-initial cause and effect.

So, now we have the basis for the conservation laws. I always wondered about that.

Yes, it is that all must sum to zero, with infinite precision, even.

What is the near 'nothing' exactly?

There is no 'exactly', due to chaos and fluctuation. It precedes real form, space, the quantum result, and probably even laws; however, it could be considered as the quantum realm, in which the near 'nothing' is the quantum fluctuation, also called quantum tunneling or quantum uncertainty.

It creates real particles, some of which are stable and enduring and capable of further combining.

Ok, but let's back up. This 'something' was always there and was never created?

True; it had to be, meaning the near 'nothing', for, that, too is a something, but I'm usually using 'something' to refer to mass-energy and matter.

Yes, but no creation means no Creator.

True. We are free.

Or, if we take Nothing as the eternal basis, then there's still no Creator.

The causeless basis, whichever it is, Nothing or near 'nothing' or 'something' must be eternal, ever its own precursor. 'God' would be complex, and so He could not be First, anyway.

And the 'How' for the mechanism forward is that anything could become of this 'chaos' of possibility, such as a universe?

Yes, for we see it in the disordered quantum realm. We even see the superpositioning of electrons in green sulfur bacteria, via fermo-lasers, as the electrons in superposition locate the most efficient path. Also, we measure atoms, each time getting a different answer. Radioactive decay just happens whenever, unpredictably. Our universe forming may have been a very low probability event, but, there is all the time in Eternity's waiting room, so, if one waits long enough, even low probability events must come to pass.

OK, then what is the 'Where' that we human mammals utilize?

Space is the 'Where'. It is a place for the 'What', which is matter.

Then whence derives our 'Then' and 'When', the past and the future, with the 'Now' ever in-between?

It's from the movement of appearances, or rotation, that being of the 'What' passing through the 'Where', which gives rise to the notion of the past, present, and future, a correspondence that is retained in memory, the now, and in imagination's projective powers.

Huh?

Remembrance, or memory, is like the past space of the 'Then' and the 'Where', whereas, history is the past matter of the 'Then' and the 'What', remembrance and history then combining into learning, using a human analogy.

And the other higher combinations?

The so-called future space is of the 'Where' and the 'When, this being like your hopes and wishes, whereas, the future matter is the 'What' and the 'When' that makes the actual progression to progress. Wishes and progression then combine into what I will call vision.

Yes, vision, but what do history and progression in combination lead on to?

A change-in-structure.

How about the result of remembrance and wishes?

They lead to a change-in-outlook.

OK, we're building up more complex shells here, but all stemming from the simpler movement of what appears; so, then, what becomes of learning combined with a change-of-outlook?

Your direction in life.

And of learning and a change-in-structure uniting?

Creating.

OK, and of a change-of-outlook and a vision?

Growth.

And of vision and a change-of-structure?

Planning.

And, finally, what altogether of direction, growing, planning, and creating?

That's your being, the 'Who'.

Alright, we've seen the beginnings reflected all the way up into our being, but what exactly led

from quarks or whatever to our much higher complexity, it taking 13 billion years?

Quarks formed into protons and neutrons, with electrons around, too, generating some the very simple elements, some of which of the preceding formed 'dust' clouds, then collected into stars, via gravity, which then generated a few more lower elements, and some middling common elements. Some stars then exploded, as supernovae, which spewed the rest of all the simple elements, as well as all of the higher elements into space, which then formed molecules that led to cells, which eventually combined into life that brought forth the brain and consciousness. That's the easy part of the TOE, for much of that is well known. Of course, planets formed near stars, and bacteria exuded oxygen into our atmosphere, as their waste product. Another few billion years and we came along, through evolution, thanks in part to the dinosaurs and 90% of all the species dying out, due to an extinction, not to mention that two proto-chimp chromosomes fused, allowing our line to branch off from the chimps and flourish, but for the times told to us by Marine Isotope 6, during our population shrunk to just a few thousand hardy souls, but we recovered from that, although it took 20,000 years.

Why did this evolution all take so long?

Death was the only chooser, plus complexity just takes time to form and grow.

So, again, no Creator, just plodding evolution.

A 'God' could have done it all in an instant.

Well, lucky us. So what the heck should we do, thrust into this life as we are, without asking?

We are truly free to make our own meaning out of life's happenstance.

I feel liberated.

Enjoy the freedom.

The Deduction of the Theory of Everything

1. There is existence; we are in it. While we are only privy to what is formed within our brains, we know there is a reality 'out there' because our senses take it in. This point is made since some say that reality is a projection; however the projector would still have a real existence.

2. The base, root existence could have no prior existence making it up, leaving only nonexistence to constitute it somehow. Either existence or nonexistence is the basis of all.

3. The basis of all, then, is either a thing or nothing, each necessarily eternal and everywhere.

4. The basis of all must be eternal, of forever duration, for it would not be the basis of all if another basis was before it.

5. The basis of all must be infinite, of everywhere's extent, for it would not be the basis of all if another basis was outside it.

6. A lack of everything (nothing) did not happen, for there is existence.

7. We cannot say that nothing begets nothing, for we do not know what the lack of anything could or couldn't do. We can say, though, that it has no 'what' (contents), no where (place), no laws, no math, no known constraints, not anything to it, etc.

8. Either (A) the basis of all is a base existent, which is necessarily a thing, or (B) the basis of all is nonexistence, a no-thing.

9. The basis of all must be the simplest state, because it wouldn't be the basis if it could be decomposed further. We do also see that composites are formed of simpler and simpler things, down unto the minuscule.

10. There can be no cause for the basis of all, since, again, then it is not the basis of all, so, it is necessarily causeless. We do see cause and effect above and beyond it, though. For the causeless, we have to find something other than cause and effect, such as an equation, perhaps.

11. For (A), the base existent thing would have to be unbreakable, as it could have no lower parts, and also unnamable, since it was never made, hav-

ing been around forever. We don't see a way that a thing could be already made and defined in its particulars without ever having been made and defined as such. This is not to say that there couldn't be many, different base existents, this being separation instead of unity. We also don't see what could have decided the total amount of instances of the base existent(s), nor why they would be workable as they are, rather than inert. This leaves the notion incomplete, and so it cannot become right until its incompleteness goes away. Incomplete notions are always wrong, if only by the virtue of their incompleteness. Having an infinite regress of smaller and smaller things does not help the incompleteness problem, but only adds to it.

12. For (B), having the basis of all being nonexistence, the base existent(s) would have to have been created from it. not being eternal in themselves, but this is not to say that they are not ever being created, as well as canceling back into nonexistence. A great support for this notion is that there is literally nothing to make anything of, that is, no other thing could contribute to the creation of the thing of the base existent(s), and so it appears inviolate, having no way around it.

13. We still don't know what a lack of anything could do, for sure, all in all, even if we propose that it is lawless, and so anything goes, but we do know for sure that existence had to come from it, as there is no other source. Evidently, the notion of a lack of anything, or nonexistence, is not what we thought it was.

14. Since the basis of all is causeless, it is the Prime Mover, using similar language, having only itself as its precursor, and requiring nothing but itself before it our outside of it.

15. We could say that 'possibility' or 'potential' wouldn't need anything before it but the same, but we don't really know what those are, but we surely know that existence is here, and thus that it was indeed possible, and perhaps even mandatory.

16. To decide (A) from (B) more work is needed. In (A) we have infinitely old base existent(s), as indestructible, and it seems we can go no further about it. For (B), we can proceed, and so if that pans out, we can forget about (A), and would have to, anyway.

17. To see if existence can happen within nonexistence, we should have both a physics/philosophic representation of it and a confirmation by observation—that there is a zero-sum-balance of opposite polarities making up nature that nullifies all of existence in the overview of totality. I have a list that I keep adding to.

For example, and this is probably not a real biggie on the list, there are opposites on the color wheel (each color not in its opposite), such as red/green, violet/yellow, and blue/orange, as well as all colors appearing in the rainbow, black being no colors, and white being all colors together. This just happens to match, respectively, Christmas colors, early spring colors, autumn colors, summer rainbows, night, and day. The list is presented after all of these points.

18. So, there seem to be balances of opposites everywhere in nature, and thus taking them all together, maybe some more than others, they might cancel out to nullify all of existence, overall, but not in practice if a lack of anything (nothing) is a perfectly unstable situation that cannot be or stay.

19. Nothing is the simplest state, so maybe it is somehow perfectly reactive. We do note that sim-

pler and simpler actual things do readily go through changes of phases and combinings and recombinings, and that higher and higher composite complexities become more stable, but not that they are everlasting or even close to it.

20. Since nothing is a lawless state, it may produce all sort of variations of things, every so often producing these in the low probability event of an entire universe, which has to still happen sometime, in Eternity's waiting room, and so it was that the mixtures in our universe were such that life could arise, although perhaps life being few and far between the billions of stars and galaxies and taking billions of years, even so, on our planet, this world also being in the right place to avoid total extinctions. The near extinctions, in which 95% of all life was extinguished actually helped us along by making openings for our forebears to evolve.

21. A nervous, shrew-like creature, upon seeing that the dinosaurs were gone, said, "Hurray! Now I can evolve!" (Not really the speaking part.)

22. So, during eternity, and throughout infinity, everything happens everywhere, even many times

over. There are many Austin's, John, and Melanie's, and always were, and ever will be.

23. Due to determinism, brought about by cause and effect, there will always be exactly what will be.

24. The same with there being no free will, for will must depend on something, which is also how we want it to be. While learning can enlarge the will, to a new fixed state, there may be some who cannot learn, and so the knowing that there is no free will causes the further learning by us to have compassion for those who may be stuck in some way or another.

25. What good is existence if we are just a kind of tourist along for the ride, which also seems that our consciousness is, too, the subconscious analysis being completed beforehand? Well, it all still appears novel to us, and enjoyable, plus we have our own meaning out of it through existence while we live through our infinitesimal parentheses in the entire Eternity, which eternity of happenings everywhere has no real meaning overall at the ultimate level. It couldn't be any other way, and we wouldn't want a certain meaning forced on us anyway. So, in a way, it is a liberation, yet the sum in-

formation content of everything is the same as of nothing, which is zero. It's like the Library of Babel that contains every possible book, even ones of gibberish. We wouldn't know which variation of a book is right, because none are right and none are wrong. There is another Library of Everything next door to the Babel Library. It is a small, empty hut, containing nothing.

24. What is life? To know the answer, one must live it fully.

25. To replace cause and effect at the base level, where anything goes as the ultimate chaos of no laws, we could use just that—the disorganized law of no laws; but the source is still Nothing, and so it ever has to amount to nothing; so, an equation is the replacement—the zero-sum balances of the sum-things produced, and this leads to the necessity for the conservation of energy, momentum, charge, baryon number, and angular momentum, and perhaps more.

26. Nothing is the Prime Mover, requiring nothing but itself. The nonexistence of Nothing is perfectly neutral and symmetrical, while existence within nonexistence must be polar and asymmetrical for it to nullify itself back to nonexistence, yet,

that doesn't happen, for Nothing is perfectly unstable It would take a God to hold it together.

27. Nothing cannot be, so something must be. There is no choice, no option, and thus no Decider.

28. If there can be one universe, then it seems that there could be others, as neither its time nor place seems special. Earth is not the center of the solar system, the sun is, and the solar system is not the center of the galaxy, but on an outer arm, far away from the maelstrom at the core of the galaxy, and our galaxy is not the center of the universe, and the universe is probably not at the center of the multiverse. And humans have only been around for a relatively short while. There are 50-80 millions species here; we are but one of them.

29. Without the moon, the Earth would have wobbled like a top. Things are just right here, such as the food chain, our few miles of livable atmosphere, a bit of fresh water, bacteria—the true Kings of the Earth, inside us, making for digestion, and much more, etc. If any important link gets disrupted, we could be doomed.

30. There are trillions of stars, and no one seems to know why. Perhaps there needs to be such immensity because the infinitesimal is so small, as a balance, both seemingly unbounded. Or perhaps we could only find ourselves in a universe that was so large that at least some small amount planets would give hope for life, even beyond the fact the the universe has the right ingredients to begin with. Because we made it, we can always look back and expect to see what some might think were fortuitous happenings, like that the Earth didn't get completely blown away by huge asteroids, or that the dinosaurs died out to pave our way, or that two chimp chromosomes fused to make new chimps that were incompatible with the old chimps; but, again, since we are here, we already know, even without looking back, that history had to go our way.

31. Evolution also selected for the notion of agency in nature, right or wrong, as a fine short cut for things having real agency, and some people carried it on into nature spirits and myths. It became better to suppose a bush rustling in the night from the wind to be of a bear's doing, for it was better to be wrong than become dead, and so we invented ghosts and more.

32. People thought simply in the old days, likening the big Myth to that of the family structure, life coming from life, with a strict father figure commanding, but they couldn't conceive that if life needed Life before it so then would Life need LIFE before it even all the more, or they didn't care, since they had gained comfort from the belief.

33. Look for higher evolved life in the future, not in the past, for that would even be at the complete wrong end of the spectrum.

34. Humans may not be well made to survive in space or on other planets, since we were fine-tuned by evolution for this world, but if we don't colonize space, then that may be the end of us, if we even make it that far, but, whatever will be will be, even to infinite precision, and that is fine, that what gets done depends on what goes before. And no one would even want all kinds of happenings going on without cause.

35. The universe is expanding, and this expansion is accelerating, unless there is some other reason for the red-shift, like photon decay, that has now sped up, and so, fairly for sure, the final fate of the universe will be to die out by so much dispersing that even one photon will not know of any other one.

36. So, the universe is ever winding down, like a spring unwinding, but this is what made for energy being able to accomplish things, it being restrained by slow and patient time so that everything didn't all happen at once in some flash of a big mess.

37. Consciousness is a fundamental property of organized matter that cannot be further reduced to elementary properties. Not all neurons are involved in consciousness; for example, not much of the vast retinal/vision system, and many other areas whose injury or impairment does not stop consciousness.

Consciousness is further seen to be of a brain process because the following can stop consciousness: Anesthesia to the brain cells; a blow to the head; sleep; too much poison/drugs. We can also probe the brain to make conscious visions appear.

38. Free will = none; it is fixed, although dynamic, via learning, and still fixed, in between, and at any instant. We believe that we have free will because of the sense of agency, the feeling that we willed an action. The feeling of agency is nothing but a conscious percept, with an Neural Consciousness Correlate that can be studied like any other. The brain makes decisions before the conscious mind is aware a decision has been made. This tells us not how decisions are made, but that decisions are made before we become aware of them and that agency plays no role in making them. The agency aspect of free will is orthogonal to how decisions are made.

The main reason why the will cannot be free is that it depends on prior things. A will that depended on nothing would be a mini first cause that had nothing to draw on. The opposite of 'determined' is 'undetermined'. We wouldn't want that even if it was an option; yet, some want free will to be so.

39. Time is distance, and distance is the difference of space(s), that is, between here and there—the movement of something. This makes space the difference of time. So, time is a difference dimension, not a compositional one. One form of time is the displacement caused by motion. Is there another form, having to do with the 4th-dimensional aspect of time? It seems that time is the dimension that bounds, not extends, 3-dimensional space.

Together, we have spacetime from space and time. Spacetime is the internal product of space and time, they seemingly woven together, but spacetime is also the product of energy and distance, for energy occupies space, or energy is space, the influences of E/M defining it.

Is space*time the same as energy*distance?

It is if energy = time*distance^2, energy spreading in time as the square of the distance. Space*time (ddd*t) = energy*distance (tdd*d).

The speed of light seems to be absolute and so is the absolute dimensional equivalent between space and time, as distance(space) and time.

Spacetime = time*distance^3, and so if c, the speed of light, has the dimensional units of of distance/time, then, solving for the external All, we have tddd * d/t = dddd, or distance^4, giving the All, which has no time, since it canceled out, a hypercube.

We might have said time*distance^3 is distance^4, if time is distance, but distance^4 seems to be an external view, and, so, internally we seem to need time*distance^3, in which relative time exists. Time may be more of a subfield. Einstein thought that something had to give, and that was time, so he didn't have it as absolute, as time being distance would have made it.

If space's three extents are defined by summation, then only one degree of freedom is left for a difference operator, namely time, to perform the utter nullification of space, since there is only nonexistence (nothing) to make anything of, and so the 4th dimension of time must have a polarity of a positive and negative axis, such as the polarity of charge that we see energy has. So, the second form of time could be charge. The 4th dimension, then, is not composed of points, as are the other three, but represents the points' 4-dimensional deflection, which is a difference of position.

c, the speed of light, underlies the dimensional relationship between time and distance and between the external hypercube view and the internal spacetime view, as,

Distance^4 = c(time*distance^3),

for c can be no only than what it is, as absolute.

Energy density would be the 4th-dimensional slope of space.

Just as Planck's constant is the 4-dimensional quantization of photons, elementary charge is the 4-dimensional quantization of matter particles.

The Cosmos in its external totality must be neutral and symmetric, to sum to nonexistence, whereas its internal composition must be polar and asymmetric, to have existence within nonexistence.

$$
\begin{array}{c}
\text{C} \\
\text{h} \\
\text{S p a c e} \\
\text{r} \\
\text{g} \\
\text{e}
\end{array}
$$

Note: Because spacetime is 4-dimensional, we have to get used to thinking of that extra dimension of time, even though our 'now' is only three-dimensional, for it has no 'time' to it, as when

time stops in movies and everything freezes in place.

Time is not just a difference of space but of spaces. Time, as any higher dimension does, touches all lower dimensions, but it also moves along as its own dimension, so, somehow, think of many 3D-spaces stacked up into that next dimension, like a stack of infinite 3D-spaces pancakes with no height limitation, and that is what time is a difference of, or see time as covering two different slices of an Einstein 4D Block Universe, as a 4D 'distance'.

40. Gravity seems to be the odd-man out of the four fundamental forces, but I think it stems from all of them, in a blend. The strong and weak nuclear forces, which some call sub-nuclear, oppose each other, as the strong force promotes stability, while the weak force promotes changeability. It's a fine situation because then things are not so frozen that they cannot change, and not so wildly changeable that things cannot stick together. There are many such balances in nature. The electric and magnetic forces are not oppositional but completely two-way transitional, each giving rise to the other in turn, in a self-regenerating wave. Gravity is still AWOL, apparently not having a

way to fit into this fine oppositional/transition scheme; however, what if it was the effect of all the forces together?

41. Perhaps from the driving apart of virtual particles by inflation faster than they could cancel out, the universe was born. The the basic particles formed stars, which produced some more of the (lower) atomic elements, supernovae spewing out the rest, which atoms formed into molecules, plants, and then into cells, and life, in some of which consciousness emerged, which allowed us to actionize without moving, and also know about what was going on around us and in us.

We became—because stars died. We are the universe come to life. It only took 13-14 billions years. Cosmic and biological evolution is one really slow dog.

42. Neither "from nothing" or "stuff having been forever" is 'god', even being the furthest from it, and, also, since either situation is eternal, there was no creation of the situation, and thus no Creator.

Recap:

The basis of All can only be the simplest possible state. Higher things cannot exist until the parts are put together. This includes beings, and so they cannot be the basis of All, fundamental and absolute, etc., as the original basis and giving rise to all else. This basis could have had no creation, for then it wouldn't have been the basis at all.

"Consciousness making all" could not be the case, for then light would not be mostly cut off every time one closes their eyes, since, then, all should really continue on, via consciousness; however, the blind cannot see and the deaf cannot hear. We have senses because there is something "out there" to take in.

Either (1) the basis of All is Nothing, since there is nothing to make anything of, or (2) the basis of all is an eternal thing, because something cannot be made from Nothing. There you have it, so, the true TOE is contained in one of the above; thus, we have localized the TOE, which is amazing in itself.

The basis cannot be composite, or it wouldn't be the basis, as its parts would be more basic; so, it is the simplest. Even though both (1) and (2) may seem to have problems, we know, for the right one, that these would not be real problems, but misunderstandings, for one of them must be true. A thing cannot be already forever and eternally be already made and defined without even having been made and defined, for then why is it how it is in particular? What would be the source from which the most basic thing could be made when there is no other thing to make it of? How could something arise from not anything? Something has to give, and at least we already know that, for one of the cases must be true. Perhaps not just one kind of basic thing comes to be, but all variations there of, and some of these recipes work and some don't. By way of empirical observation we note that higher things come from simpler and simpler things, and that there is a balance of opposites making up nature.

If the basic things were made, there is literally nothing to make the basics of. This is something that we know is true long before its proof, given the 'if'. So, one is inclined to think that the basics are and were eternal, unless Nothing can make things.

Apparently, a lack of anything (nothing) would be a completely and perfectly unstable situation. To confirm, we look at simpler and simpler things and see that they are less and less stable, readily changing, recombining, and/or going through phase changes, even to the point of popping back out of existence. Nothing is the simplest state, then, if it could even get close to being, but it cannot, making the near-nothing of the quantum fluctuations/uncertainty/tunneling the simplest state that can be and stay.

Energy or substance forever being so has a problem, which is that there would have been no point at which its total amount could have been determined by being decided. Why not a bit more or a bit less of it?

QED: Nothing cannot be, so something is, but it must balance to Nothing, but only in the overall view.

Each proposition gives a little: sum-things ever arise, and always did, forever, as pairs of opposites, constituting a zero-balance overall. Existence is of nonexistence; they are one and the same.

This All has no limitation of extent or duration; else it would not be the All if something was before it or outside it. Everything happens, everywhere and always. Stuff is forever, but it is not the same exact forever enduring stuff. Nothing still cannot be, yet it still plays a role in the Why, What, and How of all things.

The zero-balance necessity begets the conservation laws of energy, momentum, and angular momentum—of point-of-view invariance, as discovered by Noether.

There is action and reaction; conservation cannot be violated. Every particle is exactly at the place it should be at, and can not be even an iota off of its mark, as well as its energy-mass, etc. All credits and debits must sum to zero, to infinite precision. There is no skimming off the top.

The list of balances:

1. The positive kinetic energy of stuff vs the negative potential energy of gravity.

2. Positive vs negative polarity of electric charge.

3. Matter vs antimatter.

4. Everything vs nothing, each holding the same information content.

5. Fields of space vs particles in space, fields making particles maybe, and perhaps particles making fields.

6. The largest infinity vs the smallest infinitesimal, with our finite reality at the mid-point.

7. The eternal future vs the eternal past, with our 'now' at the mid-point.

8. The strong nuclear force vs the weak nuclear force, the strong for stability, the weak for change-ability.

9. Light making matter vs matter making light, each requiring the other to be previous.

10. Stellar ignition perhaps requires previous star material.

11. Electric force transforming into magnetic force into electric force, etc., as a self-regenerating wave

12. 'Now' becoming 'past' and transforming into 'future' via movement of matter through space.

13. Standing waves going both inward and outward at the same time, if they do.

14. Compression to nothing vs dispersion to nothing.

15. Positive vs negative curvatures of space, if there be such.

16. Virtual particles popping in and out of existence, always in pairs, with not enough energy to create them, to boot.

17. Two and only two stable charged matter particles in free space, the electron and the proton, and no uncharged matter particles. Only one stable energy particle in free space, the photon, neutral (or both positive and negative together), and no charged energy particles.

18. Color wheel opposites.

19. Male/female.

20. Mass/energy transition.

21. Wave/particle transition.

97. General efficiency, such as only three primary colors making up all the rest.

98. All oppositional-transitional schemes joining, such as the 4 fundamental forces having the strong vs weak in opposition and the electric to magnetic in transition, being having space vs matter in opposition and past to future in transition.

99. On/off, here/there, up/down, and all that kind of stuff.

The View from Outside of Totality

No variation

No change

This page is blank,
except for the above, etc.

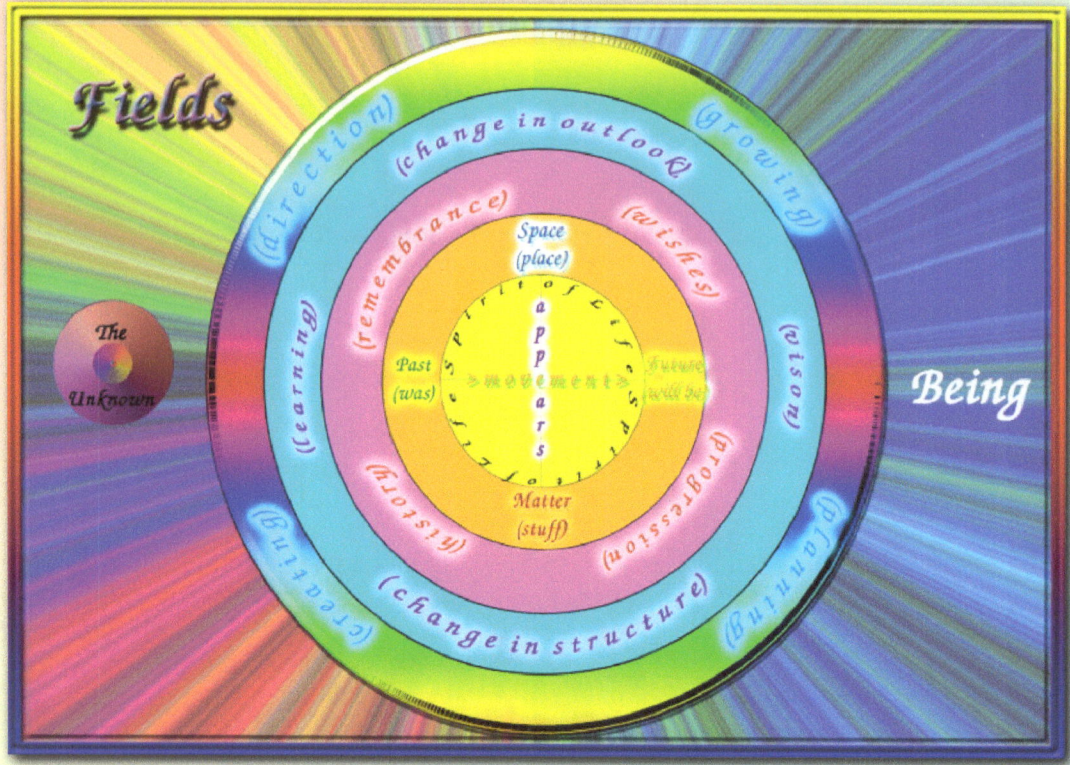

'BEING' EXPLAINED

Opposition-Plus-Transition

We often note the apparently natural scheme of "opposition-plus-transition" operating at various levels of the universe. The 'opposition', or opposing events or actions, constitute a balance of opposites, while the 'transition' is what transforms, even each into the other, sometimes.

Here is an example for the 'forces, the first one of the so-called 'pyramids':

The Forces

Strong vs. weak opposition,
plus the magnetic to/from electric transition(s).

The strong nuclear force promotes stability, while the weak nuclear force promotes change-ability. Their balance promotes progress, for events have to be orderly enough to take form, but not so much frozen that they cannot change.

The magnetic and the electric transition each into the other, in the self-regenerating e/m wave.

Gravity, a secondary force/effect, would then become of the blending of the 4 primary forces.

Fields

Universe
(regularity)

Material-
ization

Forces

Strong
(stability)

E
(motion)

Gravity blended core

condense

Gravity blended core

M
(action)

Weak
(decay)

(changability)

Evolution

Forces

S
E → M
W

[Motion]

[Matter]

Strong
(stability)

Gravity
(the blended core)

Magnetism

Electricity

Electroweak

Unknowable

Weak (change)

[Materialization]

Being

Space vs. matter opposition,
plus the past to future transition.

While space and matter are indeed similar, in some ways, the thinness of space versus the lumps of matter is a very large contrast, thus granting us a clear notion of the

'What' that is in the 'Where',
as opposed.

As for Time, it could go backwards, but it usually doesn't, due to there being so many more states of disorder than order; so, for practical purposes, the transition occurs only in one directions, as

past — > (now) — > future,

which grants us the notion of the passing of

'Then' into the 'When',
via the present 'now',

as transitional.

Time is, more or less, energy—the movement of the appearances of matter through space, or a difference of spaces(s).

The secondary blend of all this is the essence of life: one's being, which is the 'Who'.

The 'Why' of it all, which is not to say 'purpose', is that, since nothing could not be (stay), then something has to be, with no option or choice, yet it must amount to nil, as there can be no other mover that is prime, since there is surely nothing to make anything of.

The 'How' would be the necessary production of opposites from the simplest, unstable state: nothing, some of which could be noted as the opposite virtual pairs emitted, or, as like Hawking shows, the positive kinetic energy of stuff ever being canceled out by the negative potential energy of gravity.

There is more to this beginning opposition-transition of space vs matter and past to future, which is found by the further combining of the fields, on up, such as

space and past becoming remembrance,
matter and past becoming history,
space and future becoming wishes;
matter and future becoming progress,

with, then, further combining,
on up, etc., as one may predict…

history and progression leading
to a change-in-structure,
remembrance and wishes leading
to a change-in-outlook,
remembrance and history leading to learning,
progress and wishes leading to vision,

up through the remaining combinations,

onto learning and change-in-outlook
making for direction in life,
change-in-outlook and vision

making for growth,
learning and change-in-structure
making for creating,
and vision and change-in-structure
making for planning.

Finally, direction, growth, creating,
and planning make for being.

Thus, the necessary human condition.

Totality's Existence

Positive vs. minus polarity opposition (in the form of electric charge polarities), plus the mass to/from energy transition, and the matter creation to/from light annihilation transition.

Stable Existence

Antimatter(−) vs. matter(+) opposition, plus the photons to/from electrons/positrons transitions of light to/from matter.

Note that there are only two stable matter particles in free space, the electron(+) and the proton (+), because there are only two ways to make them, and only one stable, uncharged energy particle, the photon. A photon is neutral since its has both a positive and a negative aspect.

There are no stable uncharged matter particles in free space (the neutron decays within minutes).

There are no stable, charged energy particles in free space.

There is a symmetry here, on that probably allows for the nullification of existence at the level of Totality.

Finite Existence

Largest infinity vs. smallest infinity opposition,
plus the past eternity to
the future eternity transition.

Thus, our finite existence must lie
at the mid-point
of the infinite and the infinitesimal extent,
and in the centered now of the eternal duration.

Infinity * infinitesimal = 1 (finite unity),
just as n * 1/n = 1.

The no-origin Totality is its own precursor.
Think about that one.
No first universe;
no first anything.

The Summary of Composition of Being

There are appearances of lumps of matter (stuff) in a place of space, some of them in motion, whether the space is here on Earth, in outer space, or mentally, as internal representations of the external. While space and matter may ultimately be very similar at the microscopic level, at the functional life level they are an oppositional pair because we can clearly arrange stuff in space(s), such as furniture, or see stuff separately, such as an animal walking or a leaf falling, and so space is very much the practical opposite of stuff.

Space is the where of existence and matter is the what, this what in the where usually having a clear distinction. The motion of the what of stuff through the where of space makes for the other, transitional pair of then and when, of that stuff changing place, of the notions of past and future, of what was and what will be.

The oppositional and transitional pairs of space vs. matter and motion or change granting the time from past and future are the four founda-tional bases of the pyramid of being, with a spirit of life becoming the blend of all, which is similar in scheme to the forces pyramid in which gravity became as the blend of the weak force vs. strong force opposition and the electric to magnetic, and reverse transition.

So, all this gets incorporated into the necessary nature of a progressive being, for being must evolve to match what is, if it is to survive, meaning that being could hardly be much different than it is.

The actual rearrangement of matter to be started and completed as the future comes to pass is called a progression, one that hopefully began and is kept track of via the wishes that become of the mental projection in the space of the future, all of the preceding melding into what we can call vision, for lack of a better word.

On the other side of the future, which is the past, we have that the matter of the past constitutes history, along with our remembrance of it being as made of a mental space of the past, both history and remembrance ever combining into learning.

As for remembrance and wishes that operate in space, they in combination lead to a change in outlook that makes a difference, while the history and progression that are of matter combine into the actual changes in matter structures built or rearranged.

Vision and change in structure make for more excellent planning, while learning and change in structure provides creativity, learning and a change of outlook providing direction, and a change in outlook with vision providing for growth, all these in unison making for a fine and complete being.

All of this shows what becomes of space, matter, and time.

The Starring Roles

Protons formed a massive star,
Via gravity,
And for quite a while
It fused hydrogen into helium,
Living a long and healthy life,
But its death would be even more spectacular.

In its death throes,
This massive star goes out with a bang,
First collapsing,
Then triggering a supernova explosion
Bright enough
To drown out the light of an entire galaxy.

A shock wave of precious stellar debris
Hurtles outward into space
At tens of millions of miles per hour,
Containing the heavier elements
That will make up planets,
Form more stars, and even create life.

The Search

I'll follow every single avenue,
Whether it's brightly lit or a dark alley,
Exploring one-ways, no-ways, and dead-ends
Until cornered where the truth is hiding.

The Following Discusion Covers a Lot of Ground, Albeit it With Some Who Believe That Consciousness Is Everything

Since any physical base unit would have a certain definition, and not any other, then it seems that it then would have to have been created, else why its particular properties versus any other. The prime paradox seems to be that there is nothing to make anything of, yet there is something. Having something being around forever doesn't really help this dilemma. 'Nothing' could be perfectly unstable.

What is meant by your context of "unstable"?

That Nothing ever produces something. So, to confirm, we look at what we consider to be the basic somethings, like electrons/positrons, quarks/antiquarks, and maybe photons, finding that they always appear and disappear as and in opposite pairs, perhaps suggesting that these somethings are really sum-things, as a of balance of nothing. They always have opposing polarity of positive-negative and opposing matter-antimatter state. This has to be a great clue. We always get back to that there is nothing to make anything of, and so... of what else could it be made of.

Either sum-things are always being created, and going away, forever (not the same exact stuff around forever) or everything is kind of all at once. There cannot be 'God' as the base existent, for complexities ever have simpler components and so a complexity cannot be the ultimate base. Great complexity, in fact, is at the opposite end of the spectrum from any base existent. We may get to be like gods someday, but such a state cannot be First and Fundamental.

The All or Totality that is ever being searched for cannot have any beginning, for then it could and would not be all. So, no bounds of extent or duration for the Totality, since this would doom it from being the All, for it would have had something before it or outside of it. Infinity, eternity, everything, and nothing seem to all have to be in the package together.

No one has been able to suggest any other source of stuff to make anything of. It's not like we could just say that there was a warehouse of stuff or energy that just happened to be sitting around forever and available to be utilized. Still appears to be a paradox, though, since 'nothing' isn't there, but where else can we turn. Something has to give, and so we cannot be just saying 'no' right away, for we know there has to be an answer. We just don't exactly like the answer we are being led to by logic. There can be no paradoxes' only mis-

understandings. It is also said the positive kinetic energy of stuff is balanced out by the negative potential energy of gravity, yet another clue to a balance of opposites.

What about our universe's beginnings?

If one waits long enough, and there is all the time in the world in eternity's waiting room, then even low probability events must come to pass; thus the universe, and other ones, too, for if there can be one, then there can be another. This is not to say that some highly evolved alien being couldn't do some terra-forming or universe forming, but, again, such beings cannot be First and Fundamental bases.

So, we are led to a zero balance?

At least the zero-balance idea has some evidence of confirmation in the observation of how the elementals look like, whereas, the other speculations have none at all. Paradoxes can't be, so, still, something has to give. We just have to be open to find it. We have to use real clues, not just our wishes for how things ought to be, or how we desire them to be (we have to throw all that out of the window, even if is an expensive, stained-glass window).

Existence is directly a rearranged form of nonexistence. At least 'nothing' qualifies as eternal and infinite, being always and everywhere, so to speak, although it ever turns into stuff, which also goes away, which are the very strict requirements for a Prime Mover.

It could be that opposite polarity of charge nullifies all of existence (only in the overview, since, practically, 'nothing' cannot be), and/or that matter/antimatter and gravity vs. stuff also contributes.

We are really stuck with the fact that there is nothing to make anything of, but this seeming 'stuckness' may turn out to be the glory of the TOE! We just don't understand how 'nothing', which is a complete absence, could possibly do anything. Yes, existence trumps essence, totally, in everyday life. We remain curious, though.

No one can claim anything without some kind of observation and showing. Sometimes we begin with unsubstantiated ideas, for the next purpose of showing them. String theory is doing that, but cannot show anything yet. That there may be 10^{500} string theory solutions grants us more on the idea of multiple working universes.

Can we find Nothing anywhere?

We have looked for 'nothing' everywhere, but find it nowhere, as all seems to be fields. In a BBC series on 'Everything and Nothing', somewhere on YouTube, they show that some scientists pumped everything out of a large metal cylinder, yet, somethings still then appeared, these things popping in and out of existence, much like electrons and positrons do, in pairs.

What about personal beliefs?

Science does not work when there is only justification to one or a few minds, but must show the case externally to all, as plain as day. Some will still reject sure science due to emotional reasons even when it is plain as day, as emotion can wipe out reason.

Consciousness makes all; there is no objective reality.

Highly doubtful. For example, we have senses, and we also see that they take in data, from 'out there', but some still want it that all is really just like a dream, having it that the senses are a hoax and don't really do anything, the dream/consciousness making all, such as in a night dream. Consciousness still has some unsolved aspects, such as how the process turns the neural information of the 'bit' into the 'it' of our internal reality, but consciousness has been localized to the brain, and requires a brain.

God did it.

Emotions used for happiness are paramount, and we often make decisions based on them. Using them alone may not always work out. We may emotionally decide on a spacious house with cathedral ceilings at the expense of more square footage, for this will keep us happy and feeling uncrowded, and a bit of logic creeps in, too, saying that we can make a bedroom in the basement to accommodate one of the kids who would otherwise have to double up, perhaps a teenager who would like a room apart. In finding the TOE, though, other than the emotion of the awe of it, emotions can get in the way of the logic, and thus cause some amount of neglect to the facts versus what one really wants it to be.

What is time?

Time, as the course of existence, is the difference of space(s).

I like my beliefs.

A complete explanation must still entail the answers, explain WHY existence, and all things, are.

We cannot understand anything unless we know why there is existence. Any answers beyond this will be incomplete, and thus invariably wrong, due to their incompleteness.

There has always been a former existent to determine a latter one. In some sense, and in totality, existence is eternal.

Yes, eternal, and necessarily infinite, too. Would a road to the proof then be something like that since 'nothing' cannot be (which we must agree to), something MUST be (no choice, no option)? That would be progress. Is it satisfying? No option means no decider.

Some dissatisfaction may arise as to nothing deciding how things are. We could either say that they are the only way they can be, or that many ways can be, some of them workable, and some rather inert. Perhaps things are every way that they can be.

Finally, could some sort of information be the basis of existent things? But then how is it arranged and decoded.

This is in no way attacking logic: abandon logic and for the purposes of those above (analysis explanation and decision-making) you are abandoning your mind altogether.

Abandoning mind is common in human nature, and so that has to be dealt with; however, it may often call for a complete bypass, of those people, as the state is often intractable. Emotion has a direct pathway into consciousness, bypassing rational logic and reason, and so even if people are told this, they often cannot know.

What about space and energy?

It seems that space and energy must exist together, and only these two in only one way: It appears that space and energy have a totally codependent existence, as energy occupies space, so, both space and energy must necessarily have three dimensions, although one of energy's dimensions must be of the only other type: time, as energy's dimensions are proposed to be time*distance^2, while space's dimensions are as as distance^3. No other cubic is possible, is it?

We have just described 4D spacetime; Yes, 4D, three of distance and one of time. We will have to try to get used to 4D thinking.

As for energy's proposed dimensions of time*distance^2, think of energy radiating as the square of the distance over time, in that similar kind of picture often presented, that shows a square 2D cross-section slice of radiating energy getting larger and larger over time.

It is also, again, that 4D spacetime can only have two unique 3D infinitely large cubics, as there is only one 'time' dimension and three equivalent 'distance' dimensions. Space and energy are thus a necessity as distance^3 and time*distance^2, and this seems that it could be no other way. To complete this, we would have to show that only four dimensions are possible. And, of course, it is that there can be only two phenomenologically distinct substances, space and energy.

This is not to say that spacetime is not emergent, and that is a later topic which concerns the discrete, quantum, grainy bits at the Planck Scale, making spacetime not a smooth continuum. Someone at Fermilab is attempting to measure the small, jittering vibrations that would be there if the Planck scale is truly a quantum.

Seeking the truth is meaningless if you don't understand who is doing the seeking.

Essentially, we are the universe come to life, and so it is the universe itself that is doing the seeking.

The universe come to life implies it wasn't alive at some point: is that what your saying / believe?

It began with elemental particles, the simpler going to the more complex, via stars to some lower atomic elements, supernovae exploding to the higher atomic elements, to molecules, to cells, to life through evolution... to us, and perhaps aliens. We are the cosmos (come to life).

What about energy and space?

Energy is distributed into space at a certain, finite, average energy density, and this is also what requires energy and space to possess a comparable number of dimensions. It is also that for energy density to make sense that energy has to be a three-dimensional substance. Light's motion through space is an utterly explicit demonstration of the dimension energy has and space lacks!

So, the symmetry described requires space and energy to exist in equal universal quantities, and they must exist everywhere. Call it infinite if you like. Infinite, as well as eternal, is not unexpected, since the All would not be the All at all if it had a bound. There is no choice in this, is there?

The product of energy (time*distance^2) and motion(distance/time) is exactly space, a volume (distance^3). The only way energy can exist as additional volume in space is by a certain density of time per distance, which happen to be the proposed units of energy density. Energy density is circumstantial (variable), while the amount of space in energy must be existential.

We see then that energy/space = average energy density = 1 (unity), in existentially correct universal units. The universal ratio is finite because both are 3D quantities. Remember that only time sets space and energy apart. We see only two fundamental units of measurement: distance and time.

We must find why our dimensional realm is of the only dimensionality that is workable, if that is true.

Empty space, if it could be, which it can't, would have no energy, but this would be more like the nonexistence of energy, if you want to picture it; however, one cannot exist without the other, and so there is no empty space possible anywhere. So,

this is one reason why 'nothing' cannot be. While nothingness is the absence of all things, space is the lack of something, and that something is energy. There's probably a bit more to it, of why, but for now it is that this conjoining seems that it could be no other way at all.

Yes, I believe something MUST be, and it appears that it must infinitely in age and extent, yet I cannot conceive of how infinity may be possible in any physical context.

That is a tough one, as even the definition of infinity is "that which can never be attained". Does it help if the infinite amount of stuff is spread over infinite space?

I think that any configuration at all may be, so long as it constitutes existence. So what, then has determined ours? Perhaps as you said there is that time in eternity's waiting room for all possible things to pass, and we are at a certain distinct stage, but that leaves the question of the configuration of the periods of the states of existence. And these periods are but a concept, especially as time is infinite we cannot define these "periods" at all. So is our current concept of known exis-

tence simply by chance? How does chance really exist?

It seems that chance could not really exist, since events depends on what goes on beneath and before. No free will, either, then, and we wouldn't even want such a will that depends on nothing at all, for that would turn into a mess of randomness.

Objective reality is only a concept, as is other minds, and all that we will never experience with our senses. They therefore do not exist in some sense, but as posted on the other forum, Charles Taylor defines reality as all that we have, that has not left us, and so our conception constitutes reality and existence itself, but that which we directly sense, is determined at our discretion.

Our senses are in direct contact with the outer reality, but we are not. It probably wouldn't be useful for us to directly apprehend outer reality; it would be quite a mess of waves and fields going all around. Better to clearly see the color 'green' in our mind's eye than to have to sort out some certain wave frequencies directly, and even then it's not direct since it still has the sorting-out.

Now what?

So, we are still a bit stuck, which we hope is only temporary, for it is still the case that there is nothing to make existence of, and yet there is existence. An infinite regress of smaller and lesser things and causes doesn't really help, for there is again nothing to make any of that of. We must be missing something. Stuff having been forever is also a problem since it is already made and defined without ever having been made and defined. Meanwhile... We can still derive some profound implications. If existence of all has to be so, MUST BE... then it had no choice, no option... and so, no Decider. We are, then, free to be.

I've known there is no free will for a long time. I happened upon it on a trivial english question which was referring to fate... and I thought, "fate must exist, as everything is strictly defined, and predetermined..." not through this, but it seemed incredibly obvious, and so we are just living out the lives that have been determined by our nature and how the nature of the external affects it.

The will is dynamic and ever changing, but the will of any given instant is, of course, fixed to what one has become up to that point. As wider learning makes for a wider will, we can do that. For those who can't learn well or much at all, we have compassion, and even more compassion since we know there is no free will.

What we are after now is why existence has been determined the exact way it has, when there is an unlimited number of possibilities that would still constitute universal existence, assuming infinite physical existence is plausible. Because all of existence is determined by the original, eternal existents, and their nature defines and determines all. What we need is, what determines it, and what determined the beginning of, not existence, but action?

There would have to be, then, any and all kinds of existents, not just some special one, for then we would be back to why only one particular type had certain, specific properties, such as its total amount, and its own qualities of size, charge, mass, and all that kind of stuff, rather than any other definition, plus that there never was a "first place" for it to be specified, since eternal. Also, since an ultimate cause seems to be out of the question, we might have to turn to something like an equation of a zero-balance, but, as said and thought, nothing can't do anything. We are still stuck. 'Causeless' is the best answer so far as to the ultimate TOE.

There may be no such thing as time in the overview of Totality, and, if so, that reduces it to just being there, versus not there. It is, rather than is not. The so-called 'timer' of action, at least for the blossoming of our universe, could be that low-probability events must still come to pass eventually, even if it takes a trillion years (like flipping a coin to get a thousand 'heads' in a row), and so our universe was born of this. There was maybe some kind of chain reaction effect, too.

Any good clues?

The stable particles in free space exhibit a curious symmetry... Two, and only two stable matter particles, the electron and the proton, oppositely charged and with opposite mater state. (Same with their anti-particles—the positron and the anti-proton) No stable neutral matter particle (the neutron decays within minutes) One stable energy particle—the photon, neutral charge. No charged energy particles. It could be that opposite matter particles each represent half of existence, and the energy particle all of existence(positive and negative charge somehow living in peace in the photon and becoming neutral).

Deduction of infinite large and small: There are trillions of stars out there and no one knows why. This gets us to thinking how tremendously large the Cosmos is. Why does it have to be be so large? Perhaps because the infinitesimal is so small. There must be both infinite largeness and infinite smallness. It remains to be seen if the Planck size is a true limit to the small.

The final derivation: Eternal systems must be their own precursors. No first anything. All that is required is already there.

God?

Why would the eternal existents (stuff/physics) have spectral/emotional qualities like those of beings?

It's a good enough basis for the ontological argument hahaaa.. Well why do the emotions exist in themselves…? Or more specifically, how? It is a sensible possibility I believe, that some sort of universal conceptional basis pervades all that has ever been, which is tied to the view of God, etc. Though the emotions and thoughts are severely limited in rationality, feelings remain as the dominant influence, good>bad, the point of existence from the individual standpoint is this seemingly imaginary view of "goodness".

Emotions are complex, largely automated programs of actions concocted by evolution that are carried out in our bodies, such as facial expressions, postures, changes in organs, and changes in internal settings and environment. So, emotions are actions accompanied by ideas and certain modes of thinking, while feelings, from emotions, are mostly perceptions of what our bodies do during the emoting, along with perception of our own state of mind during that same period of time. So it is, that, as far as the body is concerned, that feelings are images of actions rather than the actions themselves. Emotions can be spurious, an ill effect of brain neurotransmitters out of whack, the serotonin and dopamine levels falling, due to lack of exercise and/or nutrition, or just of one's base genetics toward depression, anxiety, and obsession. Neurotransmitters regulate brain traffic. Beings and their emotions are not a base; beings have a physical basis.

I believe that Consciousness-Without-an-Object exists, however I acknowledge that the external IS, yet conceptually. In the interests of each of us I still advocate rationality and the study of the objective, even if it is but a concept alone. But God as the source of all consciousness… Consciousness must indeed have a source, and as objectively we have realized that all in totality is eter-

nal, it would seem that consciousness in totality must be, and therefore each individual must be, as we all inhabit a separate reality through ourselves, and nothing in that sense cannot be. It is the concepts that reach us alone that limit our recollections of eternity… the volume of our capacity for thoughts then must be unlimited? Hmm..

Consciousness is a process, and it requires a brain, so it is a brain process. This process can be halted by anesthesia to the brain cells, a blow on the head, or falling asleep. Consciousness of happenings appears 200-300 milliseconds after the happening, so it is last, never first, and certainly not First and Fundamental. Latching onto the word 'consciousness' and using it all by itself as a source has no real meaning.

Recap?

So, space cannot exist in the absence of energy. Space is charged with fields, converting one of its dimensions to time, that which is used in energy. Energy moves and always has polar fields because time is an essential component of its geometry. Time is a spacial difference, a difference of space(s), and spacial difference is kind of a distance, too, but a 4th dimensional one, unlike 3D distance, for time is a closure dimension that bounds rather than extends space; it is a differ-

ence dimension, not an additive one. There are only two directions along this "time" axis, positive and negative, which are called that for convenience, but they are truly opposite.

We cannot find 'nothing'—a lack of anything—anywhere that we look, for there is field everywhere, even in the fairly empty vacuum and the space within atoms. A huge metal cylinder was drained of "everything" in it, and yet something still arose, documented somewhere in this interesting video series called Everything and Nothing:

http://www.youtube.com/watch?v=JaEBM…eature=related

In short, so far, the cosmos could not exist in the absence of charged space, and charged space cannot exist in the absence of energy, for space needs the electric fields in order to nullify the baseline reference of its reference frame. So, while matter requires the existence of antimatter to balance within nothingness, space needs both to cancel the magnitude of its own closure (to nonexistence).

It looks like motion is the one-dimensional relationship between energy and space, which is also what the speed of light represents in a pristine form. Space is associated with neutral extent while energy is associated with fourth-dimensional displacement and polarity. This existential balance between space and energy must correspond to an infinite quantity of energy (but spread into infinite space).

Dr. Dudley explains 'emotion and feeling' as 'functions of an ongoing potential for weightlessness and/or infinity—hence our relationship to infinity (God)—necessarily a God transcendent of time and space.

Oh, Dudly, it's not enough that you ignore the result of billions of years of evolution, although you're bordering on it with the brain modules, and that you don't bother to undo any of what are the working of the brain, but in your bias you make up a Being for the cause of a being, and pronounce all about it instead. Your 'being requires Being' template fails instantly, beyond its making up, for it violates its own premise by halting as soon as it tries to get off of the ground. To carry it through properly, on would then have to have BEING, etc., beyond and beyonder.

Positing the supernatural just makes for a larger question because it doesn't really explain anything; it just pushes the answer off. Incomplete answers are invariably wrong, for answers must be complete. The proofs of the self-contradiction of

the supernatural remain sturdy, as well as that we find only the natural everywhere, the exact place that the supernatural is supposed to be, theistically speaking. It is really just a mere pronouncement to claim that the supernatural can be, one just born of a wish.

Zero proofs of the supernatural have been found, while there are a zillion proofs of the natural. The error is then compounded by preaching it, and that is the real problem. Why can't they say that it's just a theory? What are all these outright claims? Because then fewer would listen.

Science proof is not an internal view like the supernatural is, but is there for all to observe; however, the believers may not want to, for then the wish begins to wither. Many will go through all kinds of contortions and distortions to avoid this, including not debating, but still trying to say 'bias' as some kind of stand-in for the neglect. So much for the fiat of religious sites.

P.S. Dudly, please sign in. And the bomb was a dud. The glorious switchboard has one hundred trillion connections.

In Dudley's statement about the ancient paradox, 'we are created in the image of God—necessarily a God transcendent of space and time'.... I did

draw an insight. A synonym of 'image' is likeness and a synonym of likeness is similar—hence like God we, as consciousness/mind/perception, awareness and vision contain the potential to transcend space and time.

Neither love nor hate nor bias for an idea will be of any help, in the knowledge of the All, but even of much hindrance as to the full absorbency of the meaning inherent in the facts which are. Pronouncing all sorts of things about gods and invisible realms can never amount even to the tiniest hill of beans.

Dudley, emotional decisions can make one happy, in everyday life as lived, even for such as what the TOE should be, but, there, they can get in the way of the light of truth.

...the mind of God is immanently at work in upholding the universe as one huge and complex thermodynamic event.

God has not been shown.

...moves the organism toward infinity. Consciousness, this means, is fundamentally a contrast with infinity and is essentially a probabilistic state where the incipient or momentary absence

of something expected is, in a given moment, bound to an exponentially accelerating probability of weightlessness.

So long, Dudley.

I now believe that materialists have been seriously stunted in their spiritual growth by what, in my opinion, is a lack of exposure to certain spiritual truths, biblically revealed. For example, I believe that an individual becomes spiritually emasculated when the mother ("mater") overpowers the father ("pater") with respect to spiritual leadership.

Have a nice trip.

God is an invisible, transcendent Person – because of the way a focal image as contrasted with its ambient surround (what I will be referring to as "focal-ambience") is the actual process of creating and sustaining order in the brain.

Sorry, preaching is not allowed here.

...to when we were but a figment in the mind of God and time was essentially nonexistent. That is eternity (literally, without time), the realm in which God dwells.

Really, sermons are not allowed. Enjoy your invisibles, Dudley. I really can't see it.

I'm still searching, but I don't think anyone else really exists.

Search every dark alley. Bear in mind the unanswered questions and pursue: What is there to make anything of? Could Totality be bounded in either duration or extent? What do we know from science? How would a dream reality compare to a real one? Exact? Then, is a difference that makes no difference really a difference? Does a message differ, based on the messenger—the implementation? No. Search the heavens and the earth for answers. You are one of the few who doesn't neglect, but addresses my content. Thanks.

I can do the same for you by referring you to the 'holographic principle', in which only the 3rd dimension is projected from the real 2. This could be why the maximum entropy of a black hole depends on a surface area, not on a volume.

Positing a dream world does not solve or eliminate the real. It is but a brief respite, for it right away it gets down to the very real, again, of the Dreamer or the Projector mechanism responsible, and, so, there we are, again, asking where did it come from as real and how can it exist as real. This is meta, stepping back, or lateral thinking, a path that can often be missed when one is wholly occupied with the idea of life as a dream illusion.

Saying 'xyz' does it all doesn't really help, even if 'xyz' is replaced by some common but non supportable stand-alone words latched onto, such as 'consciousness' or 'Brahman'.

Austin .... How do we knock down a house of brick ..... if we are aware of it ?

Same as with a castle—undermine it. Then it falls of its own weight.

We all stem from one mind, one source, which has no more explanation it seems than existence, for those two details lie together.

It ain't necessarily so. Life is complex and the complex has lessor parts, so, the complex cannot be First and Fundamental. All that we observe is a progression from the simpler to the complex.

Look to the future for higher life, to the past for lower life. Sure, we can still talk about the mechanics of what makes existence. Still have to resolve the paradox of how there is existence, though, at the end of the day.

Life is produced through life it would seem that it is transferred, and so the individual mind is a part of a network of predecessors that they have inhabited, simultaneously.

Infinite regress of higher and Higher and HIGHER lives. This is the complete opposite of what happens.

We all stem from one mind, one source, which has no more explanation it seems than existence, for those two details lie together.

Bacteria. And they are still the Kings of the Earth (not dinosaurs), for they are still around, and we wouldn't even last 5 minutes without them. Luckily, for our evolved forms, they discarded oxygen as a waste product, but one being's garbage can be another's salvation.

Consciousness is all.

OK, conscious mind is the only portal through which we can experience subjective reality, and from this fact we are jumping right on to saying that minds are all there is/are, there being no objective reality 'out there', but nevertheless honestly entertaining the notion to see how it goes. We are not going to worry about how mind/brain is necessary to interpret waves and vibrations out there, anyway, to makes sense of it, for there is no 'out there', anyway. Senses and brains don't really do anything, but are projected as being there, for some reason. So, we are entertaining the notion that minds are all that is real, and nothing else is. Minds are made of mind-material-stuff; there is no other kind of material. There are no brains, no cells, no cars, no buildings, no grass, and no world. The mind makes them appear to be as so. All minds seem to agree agree on what structures and landforms appear to be there, as projected. Senses take nothing in, for, again, there is nothing out there. All that we seem to see and feel, etc., is but an apparition, but it works exactly the same as if it were real. Everything works perfectly and exactly as if things were out there. A part on a car goes bad and the car won't start. Even the whole history of how cars came into being and progressed is built by the mind, again, this being the same for everyone. If one is hit by a dream car then the mind makes a corresponding dream hurt. Only minds are real, because that is all one has access to. And so 'selves' could be real, too, since we experience a self, yet perhaps the mind is making that illusion appear as well. So let us still say that only minds are real; otherwise, we are writing words on water in a fog of unreality and cannot even trust our selves, as they would be fake.

So, all this is indeed nature's shortcut instead of having to have real stuff out there, as that would be quite a chore. Nature is efficient and frugal. The result is the same as if there were real things out there, though. Remember, a difference (in the implementation—the messenger) that makes no difference is totally no difference, truly, in the outcome—the message, as far as our experiences are concerned. This is like how music is produced by different mechanics of various mp3 players, or even by a live group, as the result is essentially the same.

We live in the Matrix, and that movie was a dream within a dream. Our minds made it, not directors, producers, screenwriters, prop and set people, camera men, and all of those thousand names one sees listed as the credits. We could be brains floating in vats, but there are no vats, no brains, no stars, and no universe, yet, mind is as real as real can be real. We seem subject to the happenings of experiences, but perhaps it is only a matter of time before we can direct our experiences, as

some can already do in lucid night dreams. All this consciousness and dream stuff may have come from a big and sleepy ultimate Guy called Brahman, and his wife, Bra-woman. They are the paternity and the maternity of eternity. However, I can give no reason for saying such a thing.

So, now what? Is there any paradox of existence, as of the mind? I would suggest that the same basic questions remain, no matter what we refer to as real, whether stuff out there, a projector mechanism, or even Bra-man, one of which is "What the heck is it doing there?"

Awareness is the 'I' of common English language usage, as in "I feel happy". Or, since we have two selves, the higher and the lower: (to one's self) "What the hell were you thinking!" This is why we talk to ourselves.

Bacteria would not be real, even now if we see one in a microscope. They are mind projections, just like the microscope.

For now, all projections are forced upon us. Brahman has many screenwriters writing soap opera scripts for us. 'I' (awareness) feel awed by all that is.

Austin, I am happy.

A feeling or sensation of happiness has surfaced on the mind from the other self of the brain, the mind only being able to handle a few things at once, and consciousness observes and witnesses this state of affairs and so one becomes aware of it as an 'I'.

Anything can exist as made by the mind, for all is in the mind.

I think I'll determine a reality of there being no dust or mess in the house to clean. Hey, I did it; it's all gone. (I cheated; I turned the lights off.)

Though physically all is a projection, we may still determine our own reality to many degrees.

That's good, for then we can make the temperature to be not of low degrees.

Austin, everything is a projection of the mind, mainly not real. All is imagined. Things exist conceptually, only in the naming of them, the mind does this.

The mind is quite a projector. Even running the Super Bowl with 100,000 spectators and all the field action is no sweat for the mind. Hey, I just

made some words appear on a screen, supposedly from some apparition calling herself Melanie.

Religion?

Religion declares its dogma all at once, which is why it comes out as the dunce. Science is of a repeatable stance, and thus brings forth reliance.

A forever basis?

The necessarily ultimate and causeless basis had to have been around forever, it being eternal, thus, it, itself, could have had no creation. This alone made the popes cry, from the pain and injury of their old dogma in stone falling, from there being no creation or Creator of the eternal causeless

Did the karma run over the dogma?

The deathly spiral of paradox ever follows

The carving of wishes into the stone hollows

Of dogma forever blocked from the allowables.

The believing dance grinds to the elemental

Of that Being who can never be fundamental.

All such tales of original stuff made of love
End where there's nothing to make it of.

A brief twilight comes and goes, as the night crashes down on them. They sit there a long time, hearing a few whales breaking through the surface and then spouting water.

Juliet: The All, meaning the TOE, or Totality, must be Infinite and Eternal, or it wouldn't be the All, and now we have Everything thrown in, as again, the All would have no limits, with Nothing somehow involved, for there's nothing to make anything of.

Patrick: Let's focus first on the Eternal, but still keeping in mind the everywhere of the Infinite.

Juliet: Eternal systems are their own precursors.

Patrick: So they must be there all at once.

Juliet: No real past, present, or future, but for how it all gets interpreted.

Patrick: No first anything, really, for forever systems.

Juliet: Light requires matter before it, but matter requires light before it.

Patrick: They were both already there.

Juliet: All at once.

Patrick: Hard to figure, but it must be so.

Juliet: Stellar ignition requires some of the elements of previous stars.

Patrick: No first star, ever.

Juliet: The All is at once.

Patrick: No electron or positron appeared before the other.

Juliet: Both at once, always popping up in pairs.

Patrick: As for quarks and anti-quarks, and even photons, which are their own antiparticles.

Juliet: Only two stable matter particles in free space, the electron and the proton, with opposing charge, and their antiparticles, of course, but no stable uncharged matter particles.

Patrick: And only one stable energy particle, the photon, uncharged.

Juliet: And from only these few all exists—in its glorious and resultant complexity, that as of now.

Patrick: Amazing.

Juliet: The All, at its level, is all past, present, and future.

Patrick: Inside it, which is no longer the ALL, time is required to traverse it.

Juliet: And time, although not fundamental in itself, always goes forward.

Patrick: There's no going back, for us.

Juliet: Why would anything at all exist?

Patrick: Because Nothing cannot.

Juliet: Yet things have no source.

Patrick: And so the total energy must be zero.

Juliet: But zero cannot be.

Patrick: So there is fluctuation, positive and negative.

Juliet: Yet that capability exists as something.

Patrick: Yes, something has to.

Juliet: Because Nothing cannot be.

Patrick: Sounds like zen.

Juliet: It grants us now and zen and when.

Patrick: Since the All is Infinite, this goes on everywhere, eternally.

Juliet: Then everything happens.

Patrick: All at once, playing out forever and everywhere, sooner or later, or even many times at once, due to infinity.

Juliet: That is the outline of the TOE.

Patrick: No first kiss for us. We've always been out there.

Juliet: And always will be.

What is the existence of the Cosmos within nonexistence, then, as there is truly nothing to make anything of? This is tough, so let's dance around it, for now.

The only way for the Cosmos to exist is for it to be as large as it is, which is infinite, because the vacuum (nothing) must be infinite, as well as eternal, again because the vacuum is so. One might also say that the Cosmos is so large because the infinitesimal of the Planck size and within it must be so small. By the vacuum we mean nothing, which is nonexistence, which may be a perfectly unstable state that cannot ever be or remain as such. There is no other possible source for existence.

The Cosmos exists now (duh), and nothingness can not prevail, and so there is a reason why that is the case. What is it? As a clue, note that no violation of energy conservation has ever been observed. What must the total amount of energy sum to?

We can only comprehend the cosmos after we know the complete answer of why it exists. Incomplete answers will not do.

Beginnings are an anathema to theories of the All, for then there would have been something else before, and so are bounded extents a problem, for then there would have been something beyond. So, there can be no first origins. There was not just nothing for a while and then suddenly an instantaneous something occurring. The cosmos had no origin and has existed forever; however, there are still trillions of stars out there and no one knows why.

If the cosmos had no origin then why is it here instead of nothing? How is existence derived from nonexistence? This is the prime paradox. No wonder everyone went off to do something else, yet, there must be a solution, for the cosmos is indeed here. The real paradox may be that something and nothing are perceived to be radically and irreconcilably different.

The cosmos' size and the conservation laws are important clues. A thing exists if it has quantity of any type. The quantity of space is volume, and the presence of space allows for the distribution of energy; otherwise reality would be but a point of infinite energy density. The sum total of quantity is reality; they are inseparable.

I didn't post it, It posted itself, as and through me. It's the IT doing it, not the me. The IT as in "it is raining" or 'it is sunny'

That darn 'it'. What a controller. In "It is raining" the 'it' (and 'rain') is a shorthand for saying that the weather is such that droplets of water are coming down from the sky. 'It' stands for how things are, weather-wise, in this case. Hope some photons form the sun are reaching you.

Electrons and positrons are always created together, and it has never been observed otherwise. The sum of all material existence, not just of some particles, is zero (nonexistence), which solves the prime paradox.

Zero energy is required for the existence of the universe because nothingness is all there is to work with in the first place. Oppositely charged matter

and antimatter sum to zero in the context of the entire cosmos, and energy must always be conserved in it to infinite precision.

Some might emotionally brush off the idea of nothing being able to be a distributed something, but, really, if it's thought out, there is literally nothing else to make anything of, and there is no way around this, and, so, intuitive or not, it has to be true. There is no logical alternative.

"Nothing"—the lack of anything—cannot be a stable state, or else it would still be the state. The cosmos had no origin from nothingness—it is just another expression of it. The cosmos is infinite and eternal because nothing is infinite and eternal, it everywhere having to be something as a distribution of it. There is no other alternative to existence. The cosmos is a perfect zero-sum equation.

Existence is a relationship, being that nonexistence can be a component of itself. Empty sets have a physical analog, and so must their relationship to each other.

Finiteness comes from a balance between infinite largeness and smallness. Infinity times zero = one (in one-dimensional space).

Evolution is nothing more than a theory on Darwin's part.

Unfortunately, for your thinking, your pronouncement falls dead and flat-out wrong. Evolution has even triple confirmation: fossil, embryonic, and DNA, which match one another. There can be no social control to truth and fact. Those who simply don't like it try to deny it.

Darwin and those who supported him seemed to have no ethical or moral awareness of the pessimism within their doctrines.

What is found is what it is, on its own. Only emotion can try to desperately sway the truth from what it is.

The Time Capsule

Since one million years had just passed by,

They, of the future, prepared to open, nigh,

The absolutely sealed container's prize,

Of a capsule made so carefully that it did survive

Without damage, being totally impregnable

About the Author

About the Author

Austin began writing for real around the age of forty, a respite from working as an Information Engineer in the field of Computer Science, doing programming, an art, as it turned out. He calls himself a humanist, and is one who enjoys the liberal arts, utilizing science, for it pervades every discipline. He is currently retired and lives in the mountains of Poughquag, NY, near the Appalachian Trail. He enjoys tennis, writing, fun, humor, thinking, sleeping, poetry, music, dining, travel, romance, reading, swimming, and life.

Connect with me online

email: austintorn@aol.com

Movies

See my YouTube Channel, 'austintorn'.
There are playlists there.

http://www.youtube.com/user/austintorn?ob=0&feature=results_main

Books by Austin P. Torney

Astronomical Wonders: Illustrated stories, poems and pictures of the universe, the sun, the moon, the stars and the planets.

Austin's Art: Best art, mostly from the other books.

Austin's Tennis Tips: Some humorous; tennis play, and stories. Color.

Brain Waves Illustrated: Astounding revelations of the mysteries of the mind and the universe. What is Consciousness and Awareness? Where did the Universe come from? What makes the mind operate as it does? What is Meditation? What is our Destiny? How do the senses work?

Butterflies At the Edge of Forever: Toe Questors from www.toequest.com discover the Secrets of the Universe, as well as the humorously dangerous implications that follow their possession of the Holy Grail of the genuine Theory of Everything. With the world's future hanging in the balance, they sharpen their wits and skills through the teachings of the learnéd Grand Masters. Extraordinary mixed media color photo composites of tropical and otherworldly scenes beyond compare. Many poems as well. Fun science, too. Much original humor. There is no greater quest than to know whence we came and what we are. Humorous but meaningful.

Elfin Legends: A journey through the otherworld.

Epic Thoughts: The Best Of: The deepest and most profound thoughts.

Illuminations: The End, the Beginning, and All that Lies Between.

Illustrated Sayings: Pearls of Wisdom.

Flora Symbolica Illustrated: The Lore and Legends of the Flowers. Color-illustrated stanzas of the myths, legends, and facts of the flowers.

Glorious Revelations: The complete scientific series of exploits.

Illuminated Revelations: Color-illustrated glorious and romantic scientific exploits.

Illustrated Take-Offs And Jokes: Humorous illustrations, many of which I made, with the remainder discovered on the internet, mainly those that are lists of things. Much original humor.

Last Knight's Almanac: The Adventures Following King Arthur's Demise Last Knight's Almanac celebrates life, love, and adventure, in the years following the death of King Arthur. It also serves as a backdrop on which I can draw literary portraits of many of man's favorite things, such as nature, astronomy, emotion, poetry, travel, history, fantasy, art, and so on. Arthurian legend is the main thread of the story and is also one of my favorite things, so this is important, too. What really happened in the Dark Ages? And yes, it's true, but only if you wish to believe it, that I unearthed these Chronicles from an iron box that was buried 1400 years ago under the Abbey of Glastonbury.

Living Freely: Life on the cheap and nearly free in a warm climate.

Magical Moments Series: Magical scenes composited as they really could be on a perfect day.

Magical Revelations Series: Magical revelations of nature on a perfect day. Part of a series. It wasn't easy getting birds to pose for me and waiting for a glorious sunset or sunrise, so I used Photoshop to create realities that could be.

Nostalgic Notions Illustrated: The Good Old Days: Illustrated nostalgic art with a paragraph below each. Relive the golden olden days for inspiration and remembrance.

Office Life: The Glad, The Sad, And The Ugly. Life at work-from my observations at IBM and elsewhere.

Reality Recomposed: Magical scenes that one might find on the perfect day at the right time.

Scientific Exploits: The Glorious, the Humorous, and the Serious: The "dry" formulas from science books are depended on for a large portion of our existence, and while we may stand in awe at their deeper meaning and even enjoy some lab experiments, they don't always reveal their full and complete history; however, the stories behind many scientific discoveries are usually insightful, amazing and sometimes even hilarious. Being that we are human, we can be doggedly eccentric in some of our earthly quests, and perhaps the genius of some scientists begets even more incredulous undertakings. (Wait until we look into the strangeness of Isaac Newton.)

Scientific Implications: Fantasy, Fancy, and Reality: Even more science exploits.

Scientific Explanations: And more science exploits.

Scientific Revelations: More science exploits.

Short Takes: The Stories Of Austin P. Torney: Original Comedy/Jokes, Glorious Nostalgia, Astounding Science, Thought provoking Satire/Take-offs, Gripping Short Stories, Deep Mystical musings, Self-Help, plus, the Theory of Everything deeply explored.

The Anti-Word: The Absurdity Of Religious Belief: The absurdity of religious beliefs is pointed out directly, as it must be, but there is also grace, humor, insight, style and the urge to fully live, unfettered by superstition. Long ago, people were thrown out of the tribe for not believing in the moon god, but that god is long gone, along with the gods of Mt. Olympus and others who had to be relocated so far away as to become invisible. Is our present day God any more real? Is He a leader? Is He a good role model. Should we follow his ways? Was God made in Man's image? Is it all just a myth? What do other insights have to offer? Lavishly illustrated. (Some overlap with 'The Guide to the All-Embracing realm of the Ultimate')

The Art of Love: Sayings (the art of) and illustrations (the art) of love.

Telling Stories: Original, illustrated Comedy/Jokes, Nostalgia, Astounding Science, Thought provoking Satire/Take-offs, Gripping Short Stories, Mystical musings, Self-Help, and even a deep science dialog with a cricket.

The Illustrated Rubaiyat Of Austin P. Torney: The Embrace Of The Human Condition: An extension of Omar Khayyam's Rubaiyat into modern times, illustrated on every other page. The Rubàiyàt stuck a chord in me which was already resonating to Omar's frequency, so, I wrote my own. Somehow, inexplicably, the verses came to me as I lived through all the experiences described therein, for I dared not write of any philosophies which had not been tried and proven. My quatrains, like Omar's, aim into the heart of life's dilemmas, offering simple common sense solutions. In this hectic, complicated world of ours we often forget that it is the simple things in life that are still the most enjoyable and inexpensive-for, everyone dies, but not everyone lives. The spirit of Omar's Persia-fume has reached me, across the centuries, as he had hoped it might, and has overtaken me unawares, inspiring me to live and write, in that order.

The Live Sea Scrolls Glorious color-illustrated scrolls of sayings to inspire living and to celebrate the human condition.

The Rubaiyat Of Omar Khayyam: Translated by Edward FitzGerald and Illustrated by Austin P. Torney: 114 quatrains are presented via a merge of the first four editions of the Rubaiyat of Omar Khayyam. The 80+ illustrations, all in color, are a mixture of ancient and modern styles, ranging from Austin's own nature compositions to his enhancement of engravings and drawings obtained from very old books. This publication celebrates the nearly two hundred years since Edward FitzGerald was born.

The Poems Of Austin P. Torney: Flora Symbolica: An epic poem on the lore and legends of the flowers as taken by Eve from the Garden of Eden. Color Symbols: Lore and Legends of the colors. Elfin Legends: Lore. The Rubayiat of Austin P. Torney: Extension of the Rubayiat into modern times. Brain Waves: The mysteries of the mind and the secrets of the universe. Misc. Poems: Science poems; Many more short poems.

The Rubaiyat Of Austin P. Torney: The Embrace Of The Human Condition: An extension of Omar Khayyam's Rubaiyat into modern times. The Rubàiyàt stuck a chord in me which was already resonating to Omar's frequency, so, I wrote my own. Text. 570 quatrains.

The Secrets Of The Universe: Wherefrom, Whereto, And What To Do: From Possibility, as in the quantum realm, but deeper still, came our account, via the near-nothing of Potential, where the buck stops, for it is neither substance nor nothing. How does everything work? What are the secrets of the universe and what do they tell us? To what shall we amount inside our parentheses of eternity? To know where we're going, we need to know where we've been, and what we're made of. Celebrate your good fortune! This is the large format deluxe edition, with all text illuminated by color plates.

The Illustrated Triumph of Life, Love, and Being: An Exploration of the Joys of the Human Condition through the Life of a Loving Couple Engaged in the Ultimate Relationship Across the Centuries and into the Future. Escaping from a monastery-abbey that engulfed itself in the flames of ignorance, such as the one in in the book 'The Name of the Rose', they, our ever returning cou-ple, salvage a mysterious book of quatrains that guides them through the joys and follies of the human condition as they live out its words, for the proof of all writing is to live it. So close in thought that they need not even be named at first, our couple takes a picaresque journey through the first part of the book to solve the difficulties of life as they are encountered in their travels through the forested countryside. Alive and positive, it makes you want to run right out and live. Includes the Book of Quatrains and the Journal. Many illustrations. Magical and Mystical.

The Guide to the All-Embracing Realm of the Ultimate:
Is it that with grace, humor and style that we will describe all the realms of life, as well as answer the ultimate questions, some via seriously comedic adventures? We will it.
Will good and evil clash, with a triumphant success? Indeed, and in the deeds of grand stimulation and logic.
Is the cause of the universe itself causeless? Yes, and we will learn why it is uncaused, just be-cause.
Did the laws of the universe come from Nothing? Just about, but there is some further ado about this near 'nothing'.
Does everything amount to a total Nothing except for the quantum fluctuations of uncertainty? Certainly.
Will we disprove the Supernatural? Naturally.
Will we hear tales of DIA events 'that never happened'? Yes, but please keep them under your hat.
Is it that we are but a shimmering glitter in the eye of eternity, a small parentheses enclosing a dust mote of a rare and lucky event of little significance on the faltering edge of forever… or, that, as our luck has never failed,

our joy and innocence will ever prevail? Who the heck knows!
What does the sum total of the information content of Everything add up to? Nothing. Nil. Null. Not a thing.
Will the Earth be destroyed in this story? Yes, but only temporarily and only a few times.
Are these your poems and art compositions? Yes. *I, Why?* Note that this is not even the shortest poem.
Did you read 'The Hitchhiker's Guide to the Galaxy?' No, I hadn't, amazingly, but when someone told me that my book was like that, I quickly read the first 'Hitchhiker' book of the series. We don't really go to many other planets in this book, but there is much earthly imagination and excitement and comedy within.
Do we really learn Everything here, such as what is the origin of the universe and also the explanation of that which produced it. Yes. *Really?* Yes, for sure.
What is www.ToeQuest.com? It's a web site where people discuss Everything. It's fun; come on over and post.
Who are your favorite authors? Victor Stenger, Lee Child, Sam Harris, Richard Dawkins, Christopher Hitchens, Michael Shermer, Jonah Lehrer, Bill Bryson, Omar Khayyam, Percy Shelley, Dan Brown, Robert Ardley, David Darling, Carl Sagan, Nelson DeMille and all the CIA/detective writers.
The unSupernatural section is long. Yes, and although it's quite funny, too, these considerations actually helped lead us to the Super Theory of Everything, so… Thank God!
Does the book get more serious as it goes along? Yes, seriously, for I am first showing the exhilaration of being alive.
You cite some of the ideas of great thinkers. Yes, and I hope that this spurs my readers to go out and buy their books to get the whole story.

The Good Book of the Humanist Bible: The UnWrit Book of Books Of the Unholy Scriptures And the UnVarnished Gospel Truth of Reality: The title says it all.

The Illustrated Poems Of Austin P. Torney: Illustrated poems on everything: life, nature, the universe and science.

The MPs' Tales: Two Army MP leaders pursue a seemingly routine weapons theft case in Hawaii, but find there are larger forces at work. Their mysterious Colonel eventually initiates and guides them towards ultimate maneuverings, in a drug case, and even to the underpinnings of reality itself.

The Theory of Everything: All the Way Up: Everything explained, its source, its Why and How, and then the Where, What, Then and When, leading on up to the Who of being, plus many discussions, and more.

The Universal Day: A journey through the hours, the life, the seasons, and the ages.

Wick and the Cricket: An enchanting tale and discussion of cosmology, especially concerning our freedom to act. Wick began this inspiring tale on ToeQuest, the likes of which may never be seen again, for it was the rarest of happenings unplanned. Crick, Kit, Carmel, and Label-Wench soon joined in. It takes place "somewhere/sometime", in the English countryside. It begins as Wick frees a cricket from his bamboo cage.

www.ingramcontent.com/pod-product-compliance
Lightning Source LLC
Chambersburg PA
CBHW051109180526
45172CB00002B/845